RAISING HAPPY
RABBITS

HOUSING, FEEDING, AND CARE INSTRUCTIONS
FOR YOUR RABBIT'S FIRST YEAR

BRITTANY MAY *AND* PENNY AUSLEY

Skyhorse Publishing

Disclaimer:
This book was written using knowledge that we have gleaned from our years raising rabbits. We are not veterinarians. The ideas found here are things that we currently use on our rabbits, but they may not work for you. All content is offered in good faith. Please use your own judgment when dealing with your rabbits, as we cannot foresee all possible situations. Please make sure you take any rabbit showing signs of illness to a veterinarian immediately. Small animals can hide signs of illness until they are very sick. When you are in doubt about treatment for your pet rabbit, consult a licensed veterinarian.

Skyhorse Publishing books may be purchased in bulk at special discounts for sales promotion, corporate gifts, fund-raising, or educational purposes. Special editions can also be created to specifications. For details, contact the Special Sales Department, Skyhorse Publishing, 307 West 36th Street, 11th Floor, New York, NY 10018 or info@skyhorsepublishing.com.

Skyhorse® and Skyhorse Publishing® are registered trademarks of Skyhorse Publishing, Inc.®, a Delaware corporation.

Visit our website at www.skyhorsepublishing.com.

10 9 8 7 6 5 4 3

Library of Congress Cataloging-in-Publication Data is available on file.

Cover design by Mona Lin
Cover images by Brittany May

Print ISBN: 978-1-51073-717-4
Ebook ISBN: 978-1-51073-718-1

Printed in China

Contents

Raising Happy Rabbits: An Introduction

Welcome! You are either considering getting a rabbit, already have a rabbit, or are just curious about rabbits. Our goal for this book is to help you make the decision of whether to get a rabbit, if you haven't made it already, and to help you give your rabbit the healthiest and happiest life possible! We have four rabbits, all of which were rescues. This path has been one of adventure, education, and joy! We hope you will find the same path, as you partake on this journey of raising happy rabbits! As long as I can remember, I had wanted a rabbit, but it wasn't until I became an adult that I was able to make the decision to get one for myself. Whether as an adult you are considering one for yourself, or perhaps you are contemplating one for your child, we are sure you have many questions about what caring for a rabbit may entail. There are many things to learn and consider whenever you take on the responsibility of a new pet. We will take it one step at a time, so don't let it overwhelm or discourage you. It is our hope that you find this book helpful, and that before you know it, you will be comfortable and find taking care of your rabbit fun and rewarding. We know that we have.

Whether you choose to keep your new pet rabbit indoors or outdoors, there are certain things that they all need. When deciding to purchase your first rabbit, all these things should be considered prior to bringing home bunny. A rabbit is not a starter or an easy pet to have. If you choose to raise them correctly, and you care about their well-being, you will discover they can be just as demanding, time-consuming, and expensive as any dog or cat.

However, if you approach getting a pet rabbit with a little research and preparation, you will soon discover that they can become your best pet ever. With time, attention, and patience, a deep bond can develop, and you will be friends "furever."

After choosing your new friend, you will be faced with many questions ranging from housing, litter training, proper feeding, exercise, freedom, and grooming to potential illnesses, playtime activities, learning what is normal and abnormal behavior, and much more.

Keeping rabbits happy is a delicate balance of allowing them to act like a wild rabbit, which is what they desire and what is instinctual for them, while realizing that they are domesticated rabbits, who actually have some very different needs, and could never survive on their own in the wild. While you may read about colonies of feral rabbits that have been abandoned in parks or areas around the world, their lives are usually short, disease-ridden, and miserable. This is why I am saying that they don't usually survive in the wild on their own. Allowing your rabbit as much freedom as possible, giving them physical and mental enrichment opportunities, and keeping a small garden of herbs and lettuces are just three ways that you can help them feel at home.

Again, we hope to help you make your first year with your bunny not only a success but also a fun and fulfilling experience that will bond you and your new rabbit for life.

What This Book Is and Is Not About

When doing research about getting a new rabbit, your goal should be to offer your new pet the best possible life they can have. This is accomplished by educating yourself about what to expect, what to watch for, and how to mix the most natural lifestyle possible with proper veterinary care, and to give them the best chance at a long, healthy life.

This book is not about breeding rabbits or raising meat rabbits. We will not be focusing on the common issues that breeding or meat rabbits encounter. Also, this book is not focusing on raising rabbits for profit. If your rabbits are breeding stock or meat stock, you still should aim for them to live a happy, healthy life. However, meat rabbits are usually only kept till they are three to six months old. They will not encounter many of the issues we will discuss here. Breeding stocks will have fertility issues, different dietary requirements, and many other pregnancy issues that we will not discuss. But please feel free to incorporate any part of this book that you feel may be beneficial if you are raising rabbits for either of these purposes.

Is a Rabbit Right for You?

When you think about having a pet rabbit, what do you envision? Rabbits can be a companion animal just as much as a dog or cat when they are raised and loved. If your only thought is a small hutch out in the garden, you are missing so much, and sentencing a very intelligent animal to a lifetime in jail. If you are planning a soft, cuddly stuffed-animal kind of pet, then you will be in for a surprise. Neither of these expectations will make for a happy rabbit. Although with built trust and patience, your rabbit may indeed be up for a snuggle in your lap.

It is my belief that most rabbits are happier and healthier when allowed to live indoors. This is not to say that you can't have a great outdoor setup available for a rabbit, it is just my personal opinion. Indoor rabbits are generally not exposed to the same health concerns that an outdoor rabbit will have to deal with. However, being able to add supervised, safe outdoor time can add so many benefits to your rabbit's quality of life.

Rabbits love activity, toys, eating, attention, and time to run free. Are these things you can offer? Setting up a spare room in your home is a great way to offer a rabbit the freedom they need to thrive. Free-roam rabbits, who have total house freedom, generally learn to be part of the family, just as any other pet will. Maintaining a corner set up for them to return to their litter box and their food and water source is all that is required. Just keep in mind that there are some very normal, instinctual behaviors, such as chewing or digging, that can get them in trouble if they are allowed total unsupervised access to the house.

Bunny Bits

Keeping a large cage available is a good idea in case you need to contain your rabbit in a small area for a short time. Being confined to small spaces for an extended period is stressful for them.

Can you spend time getting to know your rabbit's unique personality, with all of their own likes and dislikes? Can you learn to spend time on the floor getting down on their level, rather than always in a chair or on the couch? Believe me, it won't take too long before they jump up on the chair with you while you watch television. Sometimes, I am certain my Dwarf Lionhead, Sugar, is watching! Can you dedicate time to grocery shopping for a few extra things to make your rabbit's mealtimes extra nutritional? If they can't have full-time free-roaming access to the house or a room that is for them, can you dedicate several hours a day, at least, to allow them out of their cage to explore and exercise?

Finally, are you getting this rabbit for you or for a child? Honestly, it takes a child with a special personality to thrive with a pet rabbit. Rabbits get stressed and try to hide if there are loud noises, or people moving quickly around them. This is a reaction to the fact that they are a prey animal. They are on the bottom of the food chain, and therefore they have an inbred sense of fear. In the wild, this is the only way they have to protect themselves. They do not like to be chased, and more than likely, they don't like to be carried around. They also have a fragile skeletal system, and their legs, spine, and hips can break easily if they are dropped. Also, if they struggle to escape, they can easily break their back. They require daily attention to their litter box and regular eating routines that a child might not be able to keep up with. A pet rabbit may be a great choice for your child, but only you can know that. And as the adult, the cleaning of the cage, grooming of the rabbit, attention to their health, and feeding will ultimately fall on you. A rabbit takes time to learn to trust a person. It is not like a dog or cat, which will often be immediately pleased with its surroundings. It may take days, weeks, or months for your rabbit to bond with you and your child, or it may only take moments. Don't get discouraged, and encourage your child to sit quietly every day, allowing their rabbit to explore them and learn to trust them. The bond can and will develop, it just may not be immediate.

If after a few days, things aren't progressing with your child and the bunny, don't get discouraged. Encourage your child to get on the floor with them, and allow bunny to sniff them and explore; eventually, they may jump on your lap. So, don't give up, just do it on the bunny's terms, with a nice treat waiting for them. (This part is the hardest for children to understand.) And you never know, you might get a rabbit that loves attention and adores your child. Like I said, they each have their own unique personality, just like your children.

Rabbits really aren't that different from us. They need freedom, exercise, a healthy diet, and entertainment. If those sound like needs you can meet, then great! You are about to join a special group of people who have discovered the joy of pet rabbits.

Finding a New Pet Rabbit

Whether you are purchasing your rabbit from a pet store or a breeder, there are a couple of things you need to pay close attention to. I can't stress enough the importance of finding a reputable breeder if this is the way you choose to go. Some pet stores and many breeders don't care about their animals and do not care for them properly. Sadly, they just don't have the time or personnel to devote to it, or maybe they are only concerned with making a profit. The conditions that they were born into and raised in can be horrible, and this can lead to a lifetime of issues. I am not suggesting that these rabbits don't deserve good homes and to be rescued themselves, but there is a fine line between rescuing these rabbits and helping to fund the rabbit trade. Rabbits are quite often an impulse purchase, and many will end up either turned out to die from exposure or being abandoned at rescues. Often rabbits are purchased as a novelty for a child's Easter basket. Then, after a few weeks, the purchaser discovers that they have bitten off more than they can chew. These rabbits deserve another chance, and adopting from the rescues is a great way to help. You may end up with an older rabbit (six months+), but after working through the issues, you will have saved a life.

Adopting a Rescue

We will keep this section short. If you can adopt a rescue, please do. You will need to be even more patient with them because you have no idea what circumstances they have been in or for how long. You may have to deal with aggression, food aggression, health issues, or even a rabbit that refuses to acknowledge you are there. Be patient. Be patient. Be patient.

Bunny Bits

You may find many "rescues" online when you begin searching. Make sure you locate a rescue that puts the well-being of the animals first. The House Rabbit Society is an example of an organization that works hard to provide the best information and puts resources together to make sure you locate quality rescues, experienced local vets, and so much more. Their website is a must bookmark in my opinion (www.rabbit.org).

We currently have three rescue rabbits. Two of them have always been together, and we rescued them together. They were about eight months old when we got them. For eight months they were deprived of necessities. Because of this, we dealt with food aggression for over a year. They never believed they would be fed again. They were nasty when we got them, and it took days of grooming slowly to get them cleaned up. We kept pellets, as well as hay, for them around the clock because of the aggression, hoping they would learn. Anytime we approached with food, whether it was fresh or pellets, they would nearly attack us. They would eat as fast and as much as they could immediately. Keep in mind that it took longer than they had been alive to eventually help them overcome this fear. They were also timid, and very skittish. We basically did nothing but keep them fed, clean, and groomed their first year with us. We also took every opportunity they gave us to show them affection and begin to build trust. One day, there was a breakthrough, and all of a sudden they trusted us. No reason why, nothing was different than it had always been. One day, it just clicked with them that they were safe.

Our third rescue was much easier. She had been abandoned behind an office building. We were able to catch her when we were notified that she was there. She had ticks all over her, and had very little food and water for at least a month before we found out about her. But once she was cleaned up and had been fed for a couple days, she was fine.

As we have said before, all rabbits are different and will react differently. Rescue anytime you can, and don't give up on them.

Purchasing from a Breeder or Pet Store

If you are going to purchase from a breeder or store, here are a few things to consider.

AGE OF RABBITS

We are going to discuss rabbits and their unique needs beginning at eight weeks old, because you should not be adopting or purchasing a rabbit younger than that. It is actually illegal to sell domestic rabbits younger than eight weeks old in the United States. It is vitally important for a rabbit's health that they remain with their mother until they are eight weeks. She will continue to nurse them until they are seven weeks old, then she is also teaching them to eat. Her milk contains one of the highest concentrations of calories of any mammal's milk, and the colostrum they produce replenishes the antibodies and healthy bacteria that the babies need to survive. In addition to these feedings, the mother will teach the baby to eat its cecotropes (certain type of fecal material produced by rabbits), and it is a combination of all of these that create the correct flora in the gut of the young rabbit that is necessary to survive. Without this proper care for the correct amount of time, a young healthy-looking baby rabbit may slowly begin to get sick and eventually die because this balance in their intestines was not achieved.

Here is an approximate guide that will help you tell how old a baby rabbit is. It is hard to determine exact age because a rabbit can develop at a faster or slower pace. This is a basic guide:

Day 1–6: naked, eyes closed, ears back

Day 7: fur begins to grow

Day 10: eyes open (eyes can technically open between days 6 and 10)

Day 12: ears open

Day 18: will leave the nest to explore and begin to eat solid food (still nursing)

Day 60: fully weaned and independent

Bunny Bits

Baby rabbits are called kits, and a group of kits is a litter.

Photo Credit: Ann Scott, *A Farm Girl in the Making*

Photo Credit: Jennifer Smith, Bunny Besties

Photo Credit: Ann Scott, *A Farm Girl in the Making*

Researching Breeds

There are a lot of things to consider when deciding on a rabbit breed. Things like size and temperament are important decisions to make. Rabbits can range in sizes from the dwarfs like a Netherland Dwarf, Holland Lop, or Lionhead that can be as small as 2 pounds, up to the Flemish Giants that can reach upwards of 22 pounds. With many different breeds and variations falling in between. They can all be loving and make great pets. It will simply come down to what you prefer! So, big or small, lop or straight ears, fuzzy or slick fur, energetic or more laid back, every rabbit will have their own unique personality, and will surprise you.

Based on our opinions and research, slightly larger breeds such as the Dutch, Himalayan, or Flemish are usually a good choice for people who have children who will be petting them a lot. They are slightly larger and not as fragile as the tiny breeds. They tend to tolerate it better. The well-known Netherland Dwarf on the other hand doesn't usually do well with children. They prefer a more calm environment, and can be an excellent choice for adults.

Common Pet Rabbit Breeds*

Breed Name	Size	Temperament	Fur	Ears	Handling
Californian	Up to 12 lbs	Calm/Friendly	Soft/Short	Straight	Yes
Dutch	3.5–5.5 lbs	Calm	Unique Pattern	Straight	Yes
Flemish Giant	15–22 lbs	Calm/Smart	Short/Slick	Straight	Not Really
Harlequin	6.5–8lbs	Smart/Playful	Short/Soft	Straight	Yes
Himalayan	2.5–4.5 lbs	Playful/Smart	Short/Soft	Straight	Yes
Holland Lop	2–4 lbs	Playful	Furry/Soft	Lop	No
Lionhead	2.5–4 lbs	Friendly/Smart	Soft/Long	Straight	Yes
Mini Rex	3.5–4.5 lbs	Friendly	Velvet	Straight	Yes
Mini Lop	3–5.5 lbs	Calm/Smart	Furry/Soft	Lop	Yes, calmly
Mini Satin	3–5 lbs	Friendly/Calm	Shiny	Straight	Yes, sometimes
Polish	2.5–3.5 lbs	Friendly/Calm	Soft/Short	Straight	Yes
Sussex	Up to 7 lbs	Friendly/Greedy	Soft	Straight	Yes

*These are just our opinions on temperament and how the various breeds do with handling.

Buck or Doe?

A male rabbit is called a buck, and a female rabbit is referred to as a doe. It is very hard for someone untrained to identify male and female in young rabbits. So, if you are in a pet store, don't put too much faith in what you are told. A male rabbit's testicles generally do not descend until they are three to four months old, so once again, you cannot use this as a method to determine sex in younger rabbits. There are certain things you can carefully look for to determine the sex prior to testicle visibility. Here are two images to show the difference.

One or Two?

One hard decision to make is whether to adopt one or two rabbits at the same time. It is generally better, but not necessary, for a rabbit to have a friend to bond with. If they are young and housed together already, I would seriously consider adopting two together. It will make the transition easier on them. Once bonded, it is very difficult, unfair, and stressful to ever separate them. Be prepared to have them spayed/neutered as soon as possible, preferably around the age of six months, to avoid an unplanned litter of babies, if you have a buck and a doe. If you have two does, you will still want to consider a spay. The small-breed rabbits can reach sexual maturity as early as 3½ to 4 months! The larger breeds, such as the Flemish Giant, may not reach sexual maturity until six to eight months. If you are adopting at different times, you will have to slowly bond rabbits to each other, and it can be difficult. We will discuss this in the next couple of chapters.

If you do get a single rabbit, get them a large stuffed animal to be their snuggle friend, but be prepared to clean up stuffing if they decide to be rough with their new toy. Mine have never torn up a toy, though, so you may never have an issue.

Happy is our large rabbit, and our last rescue. While quarantined, we found her a stuffed rabbit to be her "HusBun." She did seem to enjoy snuggling with it, especially when we first got her. I still find her grooming it from time to time.

Basic Health

If you are getting a rabbit for the first time, please look around at their general living conditions. Do they have food, hay, and water available? Is their cage clean and free of any visible bugs? Are their eyes bright and clear, or are they weepy? Are their ears crust-free? Is the fur around their nose dry? Check their feet. Is the fur clean and are there any scabs or wounds from being housed on wire? Notice their poops; are they hard and round, or are they soft and mushy? Any rabbit in poor conditions deserves to be rescued, but it is going to need immediate help and a trip to the vet. Someone with experience needs to be contacted to rescue rabbits with special needs or severe health issues.

If you are bringing home a second bunny for bonding, be prepared to quarantine any new addition for at least a month to ensure that no issues develop that could be passed to your current rabbit. Also, make sure that you take them to your rabbit veterinarian and have them examined. This will decrease the risk of transmitting a disease to your current rabbit.

Bunny Bits

Most rabbits do not like to be held on their backs. Some are okay with it. Don't start out holding them like this. It takes practice to ensure they are properly supported, and if they panic, they could get hurt. In general, it is not recommended to hold rabbits on their backs.

How to hold a rabbit is one of the first things you need to learn. First impressions are important, and this is the case with your future rabbit as well. You need to be calm so they will sense this as you approach them. Try not to be scared or jumpy. They really do pick up on your anxiety and will usually respond the same way.

When handling a new rabbit in an atmosphere like a store, make sure to either be on a low table or on the floor with them. They can get spooked really easily and begin to kick. Kicking is dangerous for them because it can cause trauma to their backs or legs. This kicking can be frightening for both of you and can result in you dropping them, or them wriggling away.

Always approach a rabbit slowly, getting down on its level if possible, and speak to it in a calm, soft tone. When you are sure the rabbit is ready to be picked up, pick it up by placing a hand under the torso, a hand under the rear legs for support, and then pull them close to your body. Your bunny will feel safer and you will feel more confident. Never grab a bunny by the nape of the neck, ears, or legs! You always want to support their hindquarters, and maintain the natural curve of their spine. Never attempt to force them to sit or lie down. Their bones are fragile,

and any trauma to their spine, hips, or legs can lead to paralysis. Always be prepared to quickly and carefully lower them to the floor if you see them beginning to panic or struggle to get away. If they start to struggle they will try to leap from your arms and this could easily lead to a broken leg.

As your rabbit begins to get to know you, they will calm down some, which will make handling them easier. Some rabbits will never really like to be held, but most will learn to tolerate it if you learn to handle them correctly.

Bringing Home a Rabbit When You Have Other Pets

When you are bringing home a new rabbit, remember change is stressful. Your new rabbit is having to get used to being away from their old environment (good or bad), being introduced to a new home, and getting to know you. This much change all at once is quite enough to begin with for them. Give them time to adjust and claim a piece of their new surroundings as their own before they have to adjust to your other pets. Keep in mind that the smell of the other animals in your house will also be stressful for them. They are not accustomed to any of this yet.

Whether or not you are introducing your rabbit to a dog, cat, or another rabbit, I will always recommend that everyone have their own area for their food bowls. Aggression can develop really fast over food-related issues, so it is best to just keep them all totally separate from each other during their mealtimes. Also, be careful when your other animals are eating. A curious bunny may approach them, causing aggression to start.

There are many people who are very successful adding a rabbit to a home that already has a dog or cat living there. It may take some time, but there are some things you need to consider before attempting this. A lot of the success will depend on the personality of the dog, the rabbit, and you. Do you have time to nurture a relationship among these animals?

RABBITS AND DOGS

If you are attempting to add a rabbit to a home that already has a dog, I would first question the age of the dog and his or her temperament. Is he or she already aggressive? Does he or she chase squirrels and rabbits in the yard? Has he or she had any obedience training, and can he or she obey at least basic commands of "down" and "stay"?

If it is possible, specific obedience training would be great. Having someone who is trained to handle dogs instruct both you and your dog at this time would be priceless. Perhaps you can attend a few sessions before you get your bunny and explain to the instructor what you are attempting to do. Then, once you

Photo Credit: Chrissy Morgan, Seventh Heaven Farm

get your rabbit home, you can ask the instructor to come to your house for a few very specific lessons.

When beginning to teach these animals that they are now family and not mortal enemies, start with your rabbit safely secured in his cage and have your dog on a leash. This will allow you to focus on your dog and his reactions. If you see any aggression, even playful, pull on the leash and say a firm "no." Allow them to come nose to nose and smell each other through the cage. Do not immediately allow your rabbit to be loose in the room with your dog. You are aiming for them to be calm right now, knowing the other is there. Teach the dog to lie down next to the cage and stay. Constantly praise any calm behavior you see from your dog. This stage may take days or even weeks, but it is far better to err on the side of caution here. Going too fast is risking disaster.

Once you think they are calm around each other consistently, you can attempt to let your rabbit out, while keeping your dog on the leash. The importance of "down" and "stay" can't be overemphasized here. You are looking for the dog to remain quiet and calm while your rabbit explores the room and eventually the dog. If the dog gets overexcited or can't be controlled, remove him from the room, and try this step again on another day. This is a lesson in patience for you and your dog. Positive reinforcement of calm behavior is the key. Remember that a rabbit can be scared into shock if they are chased, charged, or worse yet, bitten. (You can read about shock on page 119.) If this happens, you will need to separate your animals and call your vet immediately. The key is to go slowly. Slower is better. It may take months for you to trust both animals; however, it may not! Some dogs do great with their rabbit siblings. It really does depend on their personalities. Even the calmest dog can have a moment where they "snap," and some large-breed dogs are just gentle giants. So, even if it appears to go great, don't leave them unattended together. I would be wary of this for many weeks/months, even if they are "buddies." It only takes one moment for the worst to happen.

Photo Credit: Amber Horton

RABBITS AND CATS

Cats and rabbits have a unique relationship. They can definitely be taught to like each other, but there are a few things that differ between training a dog versus a cat to get along with your rabbit. A cat who will be living with a rabbit needs to have their nails kept trimmed, removing the curve of the nail. Any scratch, intentional or not, can give a wound that can be extremely dangerous.

Just like with a dog, you need to introduce your cat and rabbit slowly, allowing them to get to know each other while your rabbit is safely confined in a large cage or caged area. This needs to be something the cat cannot get into or reach his paw through. Also, something large enough for your cat to see your rabbit jumping around in would be perfect. This is to allow your cat to become accustomed to the movements of your rabbit; this should reduce the hunting urge that every cat has naturally.

RABBITS AND RABBITS

It is possible and preferable to bond a rabbit with another rabbit, whether they are male or female. If you are purchasing or adopting your new rabbit, and you see that they are housed with another rabbit already, please take both of them together. A bonded pair will grieve deeply for each other, and they should not be separated. It is even preferable to keep a bonded pair together if one of them needs to go to the vet. The separation just causes too much stress. So, always notice if a pair was brought into a shelter together, or they are caged together.

In a communal pet store situation, notice if the rabbit you are interested in is snuggling with another rabbit, or is stretched out against another rabbit. This will be the prime time to get them together. They have already chosen a friend and forged an early bond in this stressful situation. The car ride home will be another stressful situation that

will deepen their bond. So, if they aren't being aggressive toward each other at all, and they are already caged together, keep them together in the car; don't separate them. They will comfort each other, and this is a great way to reinforce this bond.

Remember, every rabbit is unique, with his or her own personality. You may not have any issues bonding your rabbits, or it may take weeks. Don't be fooled by a great first visit together. Don't leave them alone together because they seem to be getting along great. Be careful, take the time to do the bonding correctly, or you may come into a very sad situation. Rabbits can cause considerable damage to each other in just a few unguarded moments. Once they begin fighting, it is very hard to quickly separate them. If they are getting along great, you can allow supervised playtime together, but cage the new rabbit separately in the area with your original rabbit any time you aren't there to supervise for the first several weeks.

Bunny Bits

It is preferable for your bunny to have a bunny friend! They will be happier!

If you are purchasing/adopting them separately, or at different times, the game just changed. If you are bringing another rabbit home, you will need to remember that your first rabbit owns your house. He considers you, your house, and everything in it as his territory. So, you have decided to bring in an intruder, you traitor!

Bunny Bits

The easiest situation for bonding will be if both rabbits have been spayed or neutered. This diminishes the hormonal influence and helps with the territorial drive. However, you should wait to spay/neuter until they are at least six months old.

You will need to slowly introduce them to each other, remembering this will be a very stressful situation for both rabbits, but especially your new arrival who is not only intimidated by their new surroundings but also sensing the displeasure of their newest sibling.

The ideal situation for bonding your rabbit will be a separate room that doesn't "belong" to either rabbit. Both rabbits will need to be introduced slowly; please do not rush this. Basically, you will begin by caging one or both rabbits separately and setting the cages next to each other. This gives them the chance to be near each other and get used to each other's scents. This is your basic starting point. You will need to remain at this point for days or maybe even weeks, depending on their temperaments. Eventually, you will bring the caged rabbit into the main living area with you and your "free" rabbit. This will be a signal to your new rabbit that the newbie is here to stay, not just visiting. Soon, you will be able to tell how it is going. Do they seem interested in each other? Are they aggressive at all? Do you hear a slight "humming" sound, which indicates that they are contented? Do they sit as close to each other as they can?

After this, you will begin to allow them supervised playtime together. Never leave them alone, keep a close eye, and sit on the floor with them. It is even better if you can begin by holding one of the rabbits and allowing the other to realize they are both out together. If aggression starts, immediately re-cage the new rabbit and begin the process again slowly. It will work. It just takes time. Think about it like sibling rivalry.

Once your rabbits have bonded, it will be great. They will form a deep friendship and become dependent on each other. They will love the companionship; it just takes time and patience on your part.

Bunny Breakups

It is rare, but sometimes a bond can be broken. You may never understand why, but once it is broken, it can be very hard to repair it. If you notice aggression or fighting, you must separate them immediately. A bit of time apart can definitely help cool tempers down. A "scary" car ride together can also help them realize that they need to comfort each other. I would attempt to restart the bonding process back at the beginning. If after a few months you are still seeing tension, you may need to consider a permanent separation arrangement. You can either find a way to keep them separate in a common area, perhaps relocate them to separate rooms, or lastly you can attempt to re-home one. I always hesitate to re-home because once you lose control of your rabbit, even if you give the rabbit to someone you think will be perfect, they can easily be passed from person to person and end up being abandoned or left at a shelter.

Rabbits and Guinea Pigs

It is a common misconception that rabbits and guinea pigs can easily be housed together. Honestly, they can be if you are careful and prepared for some extra work, but it is better for both of them if they have a same-species housemate. There are a few major differences in their personalities and dietary needs that make it hard to house them together. The major dietary difference is in the guinea pig's inability to synthesize vitamin C. This means that it must be supplied as part of their daily diet. This is not needed with rabbits. Rabbits are also generally larger than guinea pigs, and they can easily hurt them, whether they are being aggressive or attempting to play or even mate. Another thing to consider is that rabbits can be a carrier for the bacteria *Bordetella bronchiseptica*. This bacteria is the most common cause of respiratory diseases that develop in guinea pigs.

However, if you have a rabbit and a guinea pig already living together, and there is no sign of aggression or sickness, then don't separate them. They tend to develop very close bonds, and it could be very traumatizing to be separated.

Preparing for Your Pet Rabbit

Finding a Vet

One of the first things you need to do after adopting your new rabbit is locate a rabbit-savvy vet. Normally, you will have to find an "exotic veterinary" office to locate a vet who has any experience with rabbits. A small amount of research on the Internet should be done to locate a rabbit vet in your area, and there are some good online resources such as the Association of Exotic Mammal Veterinarians (www.aemv.org) that has a rabbit vet locator. If there aren't any in your area, you will need to find a vet who is willing to learn and help you keep your rabbit as healthy as possible. A vet like this will be willing to find an exotic vet to consult with if needed. Rabbits have a unique physiology and medical and treatment needs, and react very distinctively to anesthesia, surgery, and pain medications. For these reasons, I would reserve surgeries, unless it is an emergency, to a specialized rabbit-savvy vet, even if you have to drive a long way to get to them. Your local vet will usually be willing to handle the aftercare with the help of the specialist.

If you are still having trouble locating a good vet, then do some research online. There are several websites that have listings, by state, of vets who have been trained to take care of rabbits.

RABBIT-TRAINED VETS

While veterinary medicine has come a long way in terms of exotic and rabbit care, there are still some students who will have little if any training in the care and treatment of rabbits. During veterinary school, students may spend as little as two weeks learning about the "exotic animals." Even this training isn't rabbit specific. If a vet decides to learn more about rabbits, they have very limited options. There are some continuing education courses available, perhaps some specialized conferences. There are even some vets that learn hands-on by working with rabbit rescues and "apprenticing." As you can see, there are not that many options available for in-depth study prior to working directly with rabbits. Unfortunately, you will not want your rabbit to be the one that they learn on.

Do your research. Know your rabbit. Know their habits. Learn about their possible illnesses. Research what medications your vet should be giving you for your rabbit. Don't be afraid to ask questions.

Questions to Ask

1. Do you have any specialized training to work with rabbits?
2. If not, do you have contacts for consults with whom you are willing to work?
3. When was the last time you did a spay/neuter on a rabbit?
4. How many spay/neuters do you perform a year on rabbits?
5. What is your success rate?
6. Are you willing to prescribe medications for me to keep on hand for emergency issues with my rabbit, such as GI stasis?
7. What are your opinions and recommendations on nutrition? (This is an important part of the health and well-being of rabbits. Vets should be aware of the leading factors to a healthy and balanced high-fiber diet.)

Finding an Appropriate Cage/Hutch/Housing

This is one of the *most* important topics we are going to cover. Rabbits need space, and lots of it. Those tiny little starter packs you see in stores are just large enough to bring your rabbit home in, nothing more. The hutches that are marketed for outdoor rabbits aren't any better. Rabbits need to be free to be happy. The less time they are confined to a cage, the better. Rabbits are intelligent animals that need lots of exercise to remain healthy. Being in a small cramped cage is a jail cell for them, and will cause them to have emotional and physical issues. They can either eat too much and gain weight, or be depressed and not eat at all. They will eventually suffer issues like arthritis from having to sit in a small space so much, and will also be at higher risk of developing GI stasis. Also, a lot of cages have wire bottoms to make it easier to clean. However, it will cause them a lot of pain, and eventually it will cause sores on their feet that can become infected. If you have a wire-bottom cage you are using, make sure to provide a resting area that is large enough for the rabbit to lie down and rest from the wire.

A suggested minimum size for a rabbit enclosure is 12 square feet for their actual living quarters, with an additional 32 square feet to run and jump. Ideally, an entire room or section of a room could be dedicated for your bunny. At some point, your rabbit may be trained enough just to be a "free-range" house bunny and maneuver at will. They will prefer to be in a room with you, and this will be the best for developing a relationship with them. However, it is necessary to have an area sectioned off for their items such as the litter box, toys, food, and water, and they will like an area that they know is theirs. There are many options, but I am going to show you a couple that will provide ample room for your rabbit to be happy when they aren't free to roam.

The enclosure needs to be tall enough for your rabbit to stand on his or her hind legs and not touch the roof, and long enough for him or her to be able to jump three to four full hops. You may want to consider a slightly elevated platform that they can jump onto. This will give them a higher perspective on their surroundings. As you can see, this cannot be accomplished in many of the standard rabbit cages that are on the market.

The only exception to this rule is if this cage is going to be a part of a larger room or setup, and the door will not be closed on it. I actually use a unit like this as a transport cage, and once I get home, I can simply open the door and my rabbit hops out. They are nice to contain the "rabbit gear" such as the litter box, hay,

and water bottle. This is the only circumstance in which a unit like this should ever be recommended.

A better suggestion for indoor housing is creating your own cage that can adapt and change as their needs do. Here, you can see that office grids are used to create a huge enclosure that is full of activity toys, yummy foods, a napping area, and much more. This is a great unit to have if you must cage your rabbit during the time you are gone to school or work. They have plenty of room to bounce around and stretch. However, they'll still need free time outside of their unit. The closer you can put this to where you spend time watching television, reading, etcetera, the more time you will be able to allow your rabbit to be free-roaming, with the eventual goal of having a full free-range house bunny.

If you notice the hutch unit, it is off the ground and has a wire bottom. Normally, you find these units outside or backed up to a house or garage. Litter boxes are not needed; rabbits are not trained. They simply urinate and poop, it falls through the wire, and the farmer can rake underneath to clean up. This is not a happy life for a rabbit. Animals can get underneath their hutch and torment them. These hutches are brutally hot in the summer and cold in the winter, and usually they are miserably tiny. The wire bottom is murder on their tender feet, and this will eventually lead to infections and sores, called sore hock (or pododermatitis), which is extremely hard to cure. Rabbits housed under these circumstances are rarely allowed out to exercise. Unfortunately, these hutches are extremely popular worldwide, and people don't recognize the inherent dangers and isolated lifestyle they are subjecting their rabbits to when purchasing them.

So, instead of doing the normal hutch scenario for an outdoor rabbit, let's think about a few other options that might work out better for both you and your rabbit.

1. This outdoor area was created to keep a rabbit safe and give them lots of room to hop around while they are confined to their area. Even this isn't really enough room, and during safe times, the rabbit is allowed to roam inside another fenced area.
2. Rabbits can be housed with chickens inside a coop or barn if need be. They just need a protected area all their own. They will probably dig themselves a tunnel in the ground of the coop as a way to escape.
3. You can use a modified doghouse if you want to give them a nice fenced-in area outside.

Below are some important things to consider when building an outdoor hutch set-up. Basically, you will need to build a fortress to keep predators out and rabbits in!

1. There will need to be a safety fence all around their house. We recommend using ½-inch vinyl-coated, welded wire. This is small enough to keep out most things that attempt to break in, and the welded wire is very strong.
2. You will need to protect them from predators being able to dig under the fence. We do this by burying the same wire. Dig down at your fence at least 6 inches, all around the perimeter, and bury a 6- to 8-inch piece of fencing around the entire lot.
3. You will need to protect them from predators coming over the top. Yes, more of the ½-inch wire across the top. This will protect them from hawks, raccoons, opossums, and so much more. At the roofline, make sure the

connection is secured with screws. Raccoons can easily pry a roof off if nails are used.

4. You will need to protect them from digging out. I use the same wire on the ground under the entire floor space of the lot. A rabbit dearly loves to dig, and can easily tunnel their way out. While constructing your pen, it is very simple to dig out the entire area and lay this fencing down. You can do it all in one step if you are just starting out. After it is completely secured, put a good-quality, organic topsoil (with no fertilizers added) over the entire bottom. This will give them a cool base to play on and keep their feet off the wire.

Photo Credit: Tara Grier, Fluffy Layers

INDOORS VS. OUTDOORS

There are pros and cons for both sides of the argument about keeping pet rabbits indoors versus outdoors. It is important to remember that domesticated rabbits are not the same as their wild cousins. They don't have the ability to adapt to being outside as easily, and being confined to an outdoor hutch makes them vulnerable to a lot of issues.

Being an outdoor rabbit only benefits the rabbit if they have a safe enclosure that protects them from predators, while allowing them a large area to run and play in. The benefit of fresh air is not enough if they have to be in an off-the-ground hutch that is way too small. These hutches are also hard to situate to protect them from the hot sun of summer, and in turn they are frigid in the winter. Being outside exposes them to all sorts of illnesses that are much less likely for an indoor rabbit. Fleas, ticks, mites, and parasites are more

commonly found on outdoor rabbits. They also have a much higher probability of getting warbles (fly larvae growing under the skin of the rabbit), heatstroke, or frostbite if they live outside.

It is much better to plan to keep your rabbit safely indoors and allow them supervised time outside occasionally to play, get fresh air, run full speed, dig, and get some vitamin D. They do enjoy acting like little wild bunnies every now and then.

This is not to say that indoor rabbits can't have these same issues. During the summer you can even see indoor rabbits suffering from heat exhaustion and heatstroke, and sometimes, they can even get fleas inside, especially if you have other animals. But, generally, it is much easier to control the environment and keep your rabbits safe and happy when they are indoors.

The downside of keeping them indoors is the possible rabbit-proofing that you may have to do. Depending on your rabbit and their personality, you may have to protect cords, corners, baseboards, and even furniture from their chewing tendencies. Some rabbits never have a problem with any of this, and most rabbits get better with age, and these habits become less and less likely.

RABBIT-PROOFING YOUR AREA

A lot of rabbit-proofing is done by trial and error. You are learning each other's habits, and as your rabbit adapts to you, they may change. Chewing is a natural behavior, and a rabbit uses it to explore their world to some degree. It seems that chewing is not only necessary to keep their teeth worn down, but it can also be a nervous habit, and stress induced.

Generally speaking, it is a good idea to begin by protecting your cords. Biting through an electrical cord is annoying and expensive for you and very dangerous for rabbits. They could easily experience a strong electric shock. These tasty little cords seem irresistible to most rabbits, so save yourself a lot of money and irritation and keep them away from your bunny. You can purchase cord protectors that work perfectly to deter your new friend.

Wood furniture is either a favorite or totally ignored. You may not even need to protect it from your bunny. So, this is up to you. If you have something you are concerned with, either go ahead and protect it, or just watch your rabbit really carefully to see if they seem overly interested in it.

Baseboards are the same as wood furniture; some rabbits will dearly love to chew them, while others will never even think about it. If you have any priceless antiques or heirloom pieces in your home, please take measures to remove these items, or protect the furniture legs from your bunny's innate desire to chew them. Our Dwarf Lionhead, Sugar, loves to chew on wood, and especially loved it when he was younger. My Flemish Giant, Happy, loves to chew on fabric. She hasn't even touched a piece of wood in the house that I have noticed. I place old bedspreads or towels on the floor for her to dig and scratch about in and she loves it. Although, if there is any batting in the bedspread at all, she loves to bite it and pull it out.

Bringing Home Bunny

The day you go to pick up your bunny or bunnies will be a very exciting day for you and a scary day for them. Keep that in mind, and don't have extreme expectations about how they will act when you get them home. If you have had time to prepare their new living area, and you have everything they need ready and waiting for them, it will be much less stressful for you and your rabbit. However, we all know that impulse purchases and adoptions happen. Here are my top nine suggestions for your first few days with your new bunny.

1. The number one thing I would like to stress is for you to ask the owner/breeder/store to show you exactly what they are currently feeding your rabbit. Try to get the exact thing so that they have the same food when they get home, even if it is the cheap stuff. They may even give you a small bag of the pellets to use temporarily. You can begin working to get them on a healthier diet in a few days when they have had time to settle down. Always, always, provide them with hay and a place to hide. Please take notice of whether your rabbit was using a water bottle or a water bowl. This way you can provide the same option when you get home, whether or not you intend to keep it this way. Right now is not the time to begin teaching them something new. Perhaps they will even let you have the bottle or bowl they are currently using. You may offer both to encourage them to start drinking right away.
2. Make sure to ask if there have been any health concerns noted, or if your rabbit has ever been to a vet, and if so, which one. We would suggest having an appointment scheduled with a veterinarian experienced with rabbits within the first week for an overall health and wellness check.
3. Make sure to have a cozy towel or blanket for them to snuggle in while you are in the car. This car ride will be very stressful for them.
4. I always have a new stuffed animal waiting also. I normally try to get a rabbit toy, even though they don't know that. That way they have a "friend." If I am bringing home more than one bunny, they will each have their own friend, but I will be using the car ride to attempt to bond them to each other also.
5. If you have other animals at home, please see the section of this book about introducing your rabbit to your other pets (page 13), then proceed to the next step.
6. When you get home, slowly let them explore their new surroundings. It will be overwhelming for them at first. If you have a caged area, you can place them in there for a while so that they can eat/drink, and possibly use the litter box, or pick a corner of their cage to use first. Also, a small dish of chamomile flowers can help if they have them there to munch on. They may not even touch it

since they won't recognize it, but having it there gives them the option. This smaller area will allow them to adjust slowly. After a few hours of rest time, you can allow them out into a larger area for a limited amount of time to explore and hopefully begin bonding with you a bit. Don't be discouraged if they choose to go hide under a chair and totally ignore you. Allow them to move at their own pace, and in a little while, return them to their cage again. You are teaching them to return to their cage to eat/drink, and hopefully use the litter box. Keep in mind it takes time to litter train. Expecting accidents in the beginning will help you not overreact. You can do things to decrease a stressful environment by keeping noise at a minimum, limiting visitors, and waiting to embark on those home construction projects.

7. At the end of the first day, make sure they have everything they need in their area, including their toy friend, a hiding place, a cozy blanket, hay, water, and food. Also offer them a small treat, like a small piece of apple. (Remember, only offer them one tiny piece of a new food per day, until you see how their system adjusts to it.) Sit with them for a bit, and offer gentle head and nose rubs if they seem to tolerate it. When you finish, cover the top of their cage for the night to make them feel a bit more secure.

8. The next day will simply be a repeat. You never want to keep them caged for hours on end. They need freedom, and giving them freedom is a great way to earn their trust. If you can't be home with them all day, you need a large area for them. A cage setup is okay, as long as they have a larger area enclosed around it so they aren't confined to a tiny area all day. This will be a good way to give them space to relax and explore, while still confining them to an area while you are litter training them. You can attempt to pick them up if they aren't too scared of you, but don't force it, and don't chase them. Right now, you are simply attempting to show them they have everything they need, and that they can trust you.

9. Time and patience: be willing to give them everything they need whether or not they are acting as you expected them to. Be there, let them get to know you on their terms. Don't force anything if they struggle to get away from you. Gently set them down on the floor. They will more than likely bond to you, and I promise it will take longer than you expect. Just be patient.

Getting to Know Your Rabbit

Am I Allergic to My Bunny?

I hear this question all the time! Sometimes when you adopt a rabbit, you have no clue if you or your child may be allergic to them. Just like a dog or cat, you can actually have an allergy to your rabbit. Most of the time, doctors will immediately jump to blaming your new rabbit for the allergy, and of course, the answer is for you to remove the source of the problem. However, there are many other options rather than just attempting to re-home your precious rabbit.

Firstly, you may not be allergic to your rabbit at all. At the same time you brought your rabbit home, you also purchased a bag of timothy hay. Timothy hay has a head that flowers and produces a pollen. This pollen is a huge allergen trigger, and many, many times this hay is the source of the allergy, not the rabbit. Also, many brands of hay are very dusty, and this can also contribute to the problem. Fortunately, just because you are allergic to timothy hay doesn't mean you will be allergic to many of the other hays you can feed your rabbit. Oxbow has several different varieties, and you can purchase small bags to test your allergic reaction to them. I personally use a mixture of meadow, timothy, and orchard hay.

If you discover that it isn't the hay causing the issue, there are still a few options that can really help. Keeping your rabbit's cage and bedding clean will be a major help in reducing allergens. When you clean their area, you can use a mask to limit your exposure to the "dust bunnies." Also, make sure to wash your hands after cleaning or playing with your rabbit. Keeping your hands clean and away from your face will also help.

Purchasing an air filter and running it in the room your bunny stays in will help keep dust particles and allergens in the air at a minimum. If you aren't on allergy medications, this may be the answer to the entire problem. One of the daily nose sprays or allergy pills will make a huge difference, and you may already be taking them seasonally if you have allergy issues anyway.

Finally, I would suggest that you keep your bedroom clean and a "bunny-free zone." It may be hard, but if you keep your rabbit in specific areas, it will be much easier for you to clean and keep an air filter running than if your bunny has the run of the entire house.

Outdoor Time Ideas

Outdoor time is a controversial topic in the indoor rabbit world. There are so many things that can go wrong in an environment that you cannot control. There are predators both large and tiny. Parasites, fleas, yard plants, weed killer, neighborhood pets, hawks, wild rabbits—so many possible issues have to be factored in. However, we personally believe once most of these have been planned for, the benefits of fresh air, some sunshine, and a dig in the herb garden will outweigh the few things you can't plan for.

I never leave rabbits outside unattended. You always need to be present so that you can watch for danger, pay attention to whether they are too hot or too cold, and see what they are eating. If you are going to be out for more than ten minutes, I would take a water bowl also. You should never take an indoor rabbit outside and leave. Even if it is for an hour, they can easily have a heatstroke. They are not accustomed to the temperatures, and this switch can make them very susceptible to having an issue at seemingly mild temperatures.

HARNESSES/COLLARS/LEASHES

Taking your rabbit outside should be great fun for them. However, they should never have a collar put on them. Their fragile necks cannot support the tugging and tightness that collars will create. If you must, you can purchase a harness so that any leash is

attached to their backs, not their necks. However, if anything spooks them, they can really hurt themselves. If you do decide to go with a harness, please keep in mind that no matter how well you think you have it adjusted, if your rabbit panics, is spooked, or if he or she really wants, they can most likely wiggle themselves out of it. To get it adjusted tightly enough to ensure that this wouldn't happen would be too stressful and dangerous for your rabbit, causing not only physical harm but also mental duress. I do not advocate harness use at all. It is more desirable to have a fenced-in area for them to play in, harness-free.

PORCHES AND GARDENS

If you have a safe screened-in porch area or an enclosed/fenced-in garden area, your rabbit will have great fun feeling free to run. Keep in mind that the fence will need to be all the way to the ground, or your rabbit could easily squeeze under when you aren't looking. Playing in a rabbit-safe garden is fun, but takes some preparation. Please refer to the section on toxic plants (page 68) so that you can plan what area will be safe for them to play in.

CLEAN UP AFTERWARD

After a nice trip outside, it will be time to clean up. Keep the Huggies unscented natural baby wipes handy. Give them a good brushing, then wipe off their feet and their bottoms. Look at them really closely and make sure that you don't see anything crawling around their eyes or their bottoms. Make sure they have plenty of fresh water, pellets, and hay available once back inside.

A Rabbit's Age

The chart on page 34 is a fairly accepted human–rabbit age chart. It is definitely hard to decide how accurate this is, but during the first year of life, a rabbit will reach the equivalent of a twenty-one-year-old adult, supposedly. It is important to remember that all

data is hard to prove correct because each rabbit is an individual with their own unique traits. I believe that, just like a child, a rabbit may age physically, but they mature based on many different factors. If you have a very close bond with your bunny, and you have taught them, trained them, and they trust you, you may not have the behavioral issues that many people face during the first year.

Rabbits will reach their teenage years, and sexual maturity, around six months old. Their hormones are raging, and this is when you will notice them beginning to act aggressive and territorial. This is the point at which you should consider a spay/neuter. Any younger is just too young, and they may face many issues with the anesthesia. Doing a spay/neuter will greatly reduce these natural, but unpleasant, tendencies by removing many of the hormones that started the problem. However, if you choose not to spay/neuter, their hormones will naturally calm down between 1 year and 1.5 years old. Also, the quality and availability of daily food and exercise time will help burn out some of those hormones. A young rabbit that is caged up all day will become even more irritable, frustrated, and just be a hormonal mess. Another great reason to spay/neuter is to prevent rabbits from being able to mark their territory by spraying.

The next major milestone age you will reach with your bunny will be when they get to about eight years old. This is the point at which they technically become an "elder bun." However, you may notice them slowing down years before this. Again, they are each unique and will age differently, just like humans will. Some rabbits don't show signs of slowing down even at ten years old. Others will show a drastic change at six years old. If a rabbit has lived in cramped cages with little freedom, arthritis will likely set in earlier in life. This is another important reason to keep their environment and diet a priority.

Just watch for common issues, and be ready to adapt their environment to their changing needs. They may need a step added to help them hop onto a favorite chair, or they may all of a sudden need a diet change if something they have always eaten starts to upset their stomach.

Human Age	Rabbit Age	Human Age	Rabbit Age	Human Age	Rabbit Age
1 week	1 year	2 years	27 years	11 Years	81 Years
2 weeks	2 years	3 years	33 years	12 Years	87 Years
3 weeks	4 years	4 years	39 years	13 Years	93 Years
4 weeks	6 years	5 years	45 years	14 Years	99 Years
2 months	8 years	6 years	51 years	15 Years	105 Years

4 months	12 years	7 years	57 years	16 Years	110 Years
6 months	16 years	8 years	63 years		
9 months	18 years	9 years	69 years		
1 year	21 years	10 years	75 years		

Bunny Bits

The oldest rabbit ever recorded lived in Australia. A couple caught it in the wild, and it lived to be 18 years, 10 months old. This is according to Guinness World Records. (www.guinnessworldrecords.com/world -records/70887-oldest-rabbit-ever)

Rabbit Senses

HEARING

Sound is definitely a rabbit's most important and strongest sense. Their ears are specifically designed for their environment. The shape of their ears means that they can distinguish sounds from as far away as 2 miles. They can detect higher-pitched sounds than we can, and they use this ability to help them locate predators that they may not be able to focus on with their much weaker eyesight.

You can also tell a lot about a rabbit by looking at the position of his ears. They are able to move their ears and point them in many directions. Normally, if your rabbit has upright ears and is not a lop breed, then you can see when they begin to listen to something. Their ears will perk up straight,

and will slightly rotate in the direction of the sound. If you notice this activity, but fail to hear anything, remember they can hear far more than we can. If your rabbit is lying down, and has their ears slicked backward on their back, then they are totally comfortable and relaxed. Keep in mind that old TV sets had "rabbit ears" for antennae. It is a most apt analogy considering their job was to pick up waves traveling through the air.

Bunny Bits

It is believed by many that rabbits are able to hear and distinguish sounds from over 2 miles away! It certainly seems that way to me. I have been told that my rabbits react to the sound of my car coming, long before I pull in the driveway.

SIGHT

When rabbits are born, their eyes are closed until they are around seven to ten days old. When their eyes open, they have a unique vision that is adapted to their special needs. They have a nearly 360-degree field of vision thanks to their eyes being situated on either side of their head. An interesting fact about their vision is that they are very farsighted. Generally, they can see predators coming from great distances, but anything close appears blurry. There is a tiny blind spot directly in front of their nose, which can make it hard for them to see what they are eating. Also, it is generally believed that rabbits are color-blind, but, according to some veterinarians and scientists, it appears that they are able to differentiate blue and green at the very least. This is interesting if you think about the fact that their eyes are able to see the sky and the grass, which would be very important to a rabbit in the wild.

Albino Gene

Red-eyed rabbits are unique in the rabbit world. If they are a solid white bunny with red eyes they are an albino. If they have any coloring in their fur at all, then they are not albino, but they carry the gene. The "red eyes" are actually not red, but result from a lack of color. This means that you are actually seeing the iris reflecting the blood vessels that are running through their eyes and it makes them appear to be red. Red-eyed rabbits are very sensitive to light, so it is preferable to have them be indoors where they can be protected from direct sunlight. They also cannot see well in dim light.

Bunny Bits

Rabbits only blink 10–12 times per hour!

SMELL

A rabbit's sense of smell is as keen as their hearing. Due to their eyesight being so poor, they rely heavily on scent to distinguish everything in the world around them. Did you know that your rabbit will identify you by your smell much faster than they will ever recognize you by sight?

Their little noses seem to never stop twitching! Doing this helps them identify who is around them, detect changes in their environment, sense other approaching animals, identify food sources, locate their babies or other rabbits, and so much more.

Another feature that is unique to a rabbit is the split in their upper lip. When they are sniffing you may be able to notice that the split opens up a bit. This feature moistens the air as they are breathing it in, and this heightens the scent even more for them.

Their sense of smell is so much more defined than ours that they can detect the slightest change in a smell, much the same way that we use our eyes to detect slight changes in something. To help understand how much better they can smell than we can, we have approximately 5 million scent-receptor cells, and a rabbit has approximately 100 million scent-receptor cells.

Because of the sensitivity to smells, it is obvious why you can notice your rabbit run from you if you change your perfume or laundry detergent. They may also hide if you are cooking something that

displeases them, or perhaps you are cleaning with something new that they aren't happy about. They know what things are supposed to smell like, and sometimes they don't appreciate a change like this in their world. It is also possible that these harsh smells can even trigger an upper respiratory problem.

Bunny Bits

Rabbits are obligate nose breathers. This means that they only breathe through their noses, making any respiratory issue a major concern. If their noses clog up, they will be unable to breathe.

TASTE

I want you to think about what life would be like if every day you were only given crunchy bread and water to eat. Nothing fresh, nothing sweet or salty, just plain crunchy bread. Would you be happy? Now, many people actually believe that a rabbit only needs pellets.

Think about this for a moment. A rabbit's diet should be made up of 80 percent varied hays, and a small amount of pellets at the very least. But most people think they can buy a 50-pound bag of pellets at the store and that is all their rabbit will need for a year.

When you eat, you enjoy your food with approximately 2,000–8,000 taste buds. We can distinguish the tastes of sweet, sour, salty, and bitter. Now, imagine that you had 17,000 taste buds! Can you even imagine what food would taste like? A rabbit has approximately 17,000 taste buds that can distinguish sweet, sour, salty, and bitter. Wild rabbits can even tell the difference between toxic and nontoxic plants. Do you think any rabbit can be happy eating only cheap pellets? Would you?

We suggest a diet plan later in the book that will help your rabbit enjoy their daily food.

TOUCH

A rabbit's sense of touch is amazing. They have whiskers on their face in very strategic positions: above the eyes, cheeks, nose, and mouth. These whiskers have sensory nerve endings on each follicle, and they use these to determine whether or not they can fit into an area. Their whiskers grow to the exact width of their bodies. This means, if they stick their heads into an opening and their whiskers clear it, then their body can also fit through. This is an amazing way to protect them from getting stuck in open-

ings, tunnels, or burrows underground. This is also an important reason to never allow their whiskers to be cut. It would actually cause pain because of the nerves running through them, and it will damage their ability to determine safe areas until the whiskers have had time to regrow to the correct length again.

Rabbits have nerve endings covering their entire body, which allows the heightened sensitivity that you may observe when you touch them. They do enjoy being petted, but you have to determine where and what pressure they are okay with.

Rabbits' Teeth

Just like humans, a rabbit has very specific needs when it comes to their teeth. They have a total of 28 teeth. Unlike our teeth, their teeth have no enamel on them, which means they begin to wear down quickly with daily eating and chewing. Because of this, their teeth are "open rooted," which means they will continue to grow throughout their lives, averaging 3–5 inches of growth a year. In healthy rabbits that are maintaining a good diet, the hay that they are eating is one of the main ways that they can keep their teeth ground down by themselves. A good quality variety of hay is the best way to help keep teeth naturally ground down. Different hay types have different textures and thicknesses; this means that they chew them differently, causing their teeth to wear down more evenly. They also love to chew, and maintaining toys for them to chew on is another great way to encourage this.

However, some rabbits do have issues with their teeth even with the best care. Sometimes, their teeth grow much faster than they can maintain themselves. This is another reason a good vet is important. If a vet can trim the teeth once in a while, your bunny will be much happier.

Another issue that can happen is known as "tooth spurs." These tiny projections grow off of the sides of the back teeth, and can lacerate the tongue or cheek. This is extremely painful, and will cause your bunny to stop eating, which is very dangerous if not caught early. Once again, a trip to the vet is the solution.

Dental or oral abscesses are a dangerous condition that can develop quickly. If a tooth has an issue and gets infected, the infection can create an abscess quickly. Sometimes you will notice a swelling in the face that can signal the problem. A vet can prescribe antibiotics and pain medications, which will be needed immediately. Once the infection clears, the tooth can be x-rayed and better examined. Hopefully, the issue can be remedied easily, but if not the veterinarian can usually remove the tooth.

What to watch for with dental issues is fairly simple. Usually, the first sign will be a decrease in food consumption, a decrease in appetite, drooling, and their eyes may even start watering. Catching this early is so important because it can escalate quickly and cause many other issues. It is dangerous for a rabbit not to eat steadily. This can trigger GI stasis. Additionally, the threat of any of these issues leading to a dangerous abscess is great. So, treating early is your best bet. If you see these signs, call your vet. This isn't an issue you can wait a few days on.

Rabbits' Reproductive Tract

Sexual maturity is reached at a surprisingly young age. In small to medium-sized breeds, maturity is usually achieved at four to five months old, while larger breeds don't reach maturity until six to nine months old. You will notice a sharp change in behavior at this point because their hormones will begin to rage. You may notice a new tendency to act out mating behaviors, aggression (including nipping or biting), or that they're being territorial, even destructive. Keep in mind that they are being flooded with hormones, and they don't understand why they are acting this way, either. Thinking about them as aggressive little teenagers may be the best analogy.

This is also the most dangerous age for a rabbit to be abandoned. Most people are not willing to do one of the two things that a rabbit needs to help them through this stage of life.

SPAYING AND NEUTERING

Your first, and arguably your best, option is to find a very experienced exotic veterinarian who can safely perform a spay/neuter. In rabbits, it is also important for their health to have this surgery. Female rabbits become extremely susceptible to uterine cancer after the age of four. The cancer rates are alarming, but the spay surgery, as long as they remove both the uterus and the ovaries, will remove all risk of them developing these cancers in the future. A male rabbit also has a high risk of developing testicular cancer later in life, so it is best for them to be neutered for the same reason; however, their risk isn't as high as a females. In addition to the health benefits, your rabbit will lose those hormones that are causing the aggressive behavior, and most of those issues will slowly vanish over the next couple of weeks.

Make sure to ask your vet how often they perform this surgery and the success rate, as well as their willingness to send you home with pain medication for the next few days. Generally speaking, the most risk associated with the surgery is the anesthesia that is required. A vet who has been correctly trained and is very experienced will almost always have a great success rate.

It is important to wait until your rabbit has reached at least six months old before attempting the surgery. Some people argue that it is safe earlier than this, but I personally wouldn't want to take the risk.

This next question may tell you all you need to know about your vet's rabbit experience. Ask them if they are going to request you to withhold food prior to surgery. If the answer is yes, then I would begin to question their rabbit knowledge. Rabbits cannot vomit, and therefore there is no need to withhold food. This is actually dangerous because allowing their digestive tract to empty will run them at a huge risk of developing stasis. Be sure to ask the vet if they will be administering any pain medications after surgery. This, along with adequate hydration, temperature regulation, decreased environmental stress, and correct amount of fiber intake are important for a healthy recovery and healing process.

Another point I try to make is that I don't like to leave my rabbit alone at the veterinarian's office. I would prefer a vet that allows me to sit right with my rabbit until they are ready for surgery, and will allow me to be there once my rabbit begins to wake up. As we have discussed, they are prey animals, and if they are not used to hearing dogs barking, cats meowing, constant noise, funny animal smells, and so much more, this will add so much stress to them, especially when they are scared and hurting after surgery. I would ask if they have a quiet and warm place your rabbit can be taken to recuperate pre- and post-op. These things will significantly lower your rabbit's stress level. Also if your rabbit has a bonded mate, ask if you can bring them at the same time, just so they can remain together. The simple choice of allowing you to stay in close proximity will help your rabbit feel more secure and less abandoned when they go in for surgery.

WAITING IT OUT

Your second option, if you choose not to spay/neuter your rabbit, is to go a bit more natural. In doing this, you are taking a chance on them developing the different cancers later in life, and you may have to do an emergency spay/neuter at some point, but that will be a personal choice. By deciding not to spay/neuter, you will need to be aware of your rabbits different needs. From the onset of sexual maturity till about 12–18 months old, their hormones will be raging. By a year and a half, they seem to be able to regulate their moods, and they begin to naturally calm down. Many of the aggressive behaviors will greatly diminish or even disappear.

Bunny Bits

Keep some organic dried chamomile in a dish near your rabits' food. They will learn to love munching on this treat, and it is a natural way to calm them down some.

The number one issue you will have is the fact that you cannot allow boy/girl playtime without risking a pregnancy and/or aggressive fighting. I have successfully bonded two unaltered females with very little fighting; however, this may not always be successful.

FEMALE REPRODUCTIVE TRACT

One interesting thing about a female rabbit is that she is almost always fertile. They ovulate based on when they mate. An egg will be released after mating; rabbits do not go into a "season," meaning that their fertile times are not triggered by hormones. She is most fertile for approximately 14 out of 16 days. She will be pregnant for approximately 31 days, and can immediately get pregnant again. Depending on breed, litter sizes can range from just 2 babies up to 12.

Bunny Bits

A female rabbit can get pregnant again within hours of giving birth. This isn't a healthy option for her, but it is possible.

False Pregnancy

Sometimes in females that have not been spayed, they will exhibit a false pregnancy. In pregnant females that are close to giving birth, they will begin pulling their own fur out to create a nice warm bed for their litter.

Photo Credit: Janet Garman, Timber Creek Farm

Sometimes this begins to happen in a doe that has had no contact with a buck. This is known as a false pregnancy. She can sometimes pull so much fur from her sides and belly area that you are afraid she will be bald. You may also notice her begin to gather hay and take it to the nest. You may see your rabbit carrying a large mouthful of hay away from the hay bin; this is another sign that she is attempting to nest. Again, I start the chamomile treats, and I constantly clean up the fur. Locate where she is building her nest, and keep it cleaned out also. She will be aggressive during this time, so just be patient. Eventually, she will usually work her way out of this. This doesn't happen as often as females that have been spayed, and it is always recommended that you spay to reduce these surging hormones.

Mock-Mating Behavior

I would like to talk about a subject that may seem awkward or even embarrassing to you. It is a subject that I feel is important and a huge part of a rabbit's makeup. A rabbit, just like any other animal, is created with a natural desire to reproduce. Every rabbit is different, and this may not be an issue for you at all. I'm speaking from experience here. My Dwarf Lionhead, Sugar, has not been neutered. Therefore, he is kept apart from the does that I have, because they haven't been spayed, either. His desire to mate is very strong, so he likes to do what I call "mock-mate" or "hump" my feet. He "kisses" or licks them and even makes little nibbles on them, which is naturally what he would do with a female. The buck and doe take turns grooming each other, as part of the mating process. I touch on this subject because I feel that it is important to realize that this behavior is perfectly natural. If your rabbit spends a lot of time with you, outside confinement, this will most likely happen between you and your rabbit. "Mock-mating" is done without regard to gender. Whether you have a buck or doe, they can do this. The rabbit shouldn't be scolded or kicked when this is attempted on you. They simply will not understand, and this will affect their confidence in you and can cause them harm, not only emotionally but also physically. As previously discussed, their skeletal systems are fragile, as their bones are hollow. If you don't want your bunny to "mock-mate," gently pick him up and move him away. You may have to do this continually. I let Sugar do "his thing." It's really sweet and it doesn't bother me. If I listen closely, I can hear him "hum." He's happy and healthy, and that is what is important to me. While on the topic of "mock-mating," you need to know that your rabbit may "seize" while he's "mock-mating" your feet. This is no need for concern. I call it "seizing" because it looks like they stiffen up and can actually fall over. Usually when Sugar "seizes" he gets the hiccups and then thumps, and in no time at all, it's back to business as usual.

While Sugar is "mock-mating" my feet, we use this time for massage therapy. He will take breaks from his endeavors and lie down at my feet for a massage. I use my feet to rub his head, neck, shoulders, and body. He loves it! I'm sure he considers this as my part of the mating ritual. After he's had enough massage, he will begin licking my feet and start his ritual all over again.

I hope that you don't find this offensive. It is not my intention to offend anyone. I just feel that it is an important part of my rabbit's overall health and well-being. If Sugar had a mate, his attention would be elsewhere, I'm sure.

Rabbits' Feet

Rabbits have five toes on their front paws and four toes on their back paws. When running, rabbits generally run on their toes.

Basic care of your rabbit's feet is generally very easy. As long as there is no injury, a quick check on the length of their toenails, and a check on the bottom of their feet, is generally all they need. As long as they have plenty of freedom to run around and proper housing, their feet shouldn't pose much of a problem.

One common mistake people make is having a rabbit cage or hutch that requires the rabbit to stand on a wire floor. The bottom of their feet and heels have very little natural padding and this is why they come equipped with a very thick padding of wool on their feet. Slowly, over the course of time standing on wire, this wool padding will wear away and cause their skin to come into contact with the wire. This will cause what is known as "sore hock" (see page 115).

One thing you will quickly notice is your rabbit's aversion to hardwood or laminate flooring. They will always prefer to be on a piece of carpet. This is because they don't have pads on the bottom of their feet like cats or dogs do. These pads allow some friction when they walk. Since a rabbit doesn't have this, when they are on slick floors and try to take off running, they will slide all over the place. They have no way to get a grip on the floor.

I have area rugs and runners down throughout the house, which gives them the ability to easily get through the house. It is really cute to watch them as they learn your house. They will know what direction every rug is in, and they will learn the quickest way to get from one to another. Even if it isn't a direct path, they will hop to the next nearest rug.

Bunny Bits

Did you know that a rabbit's normal resting heart rate is between 130–170 beats per minute, but can go as high as the low 300s in times of exercise, pain, or stress, and their respiratory rate is 32–60 breaths per minute?

Rabbits' Intestinal Tracts

FEEDING

A rabbit's digestive tract is extremely sensitive, and anything can throw off the balance and create an emergency situation. Rabbits need a high-fiber, low-sugar diet, which you are going to discover is very hard to maintain for them. Rabbits have a "sweet tooth," and will beg for fruits, or cereals, or basically anything you are eating. It is your responsibility to educate yourself on what is safe, what is unsafe, and what is toxic. Also, any new foods, pellets, hay, fruits, or vegetables should be introduced one at a time, with several days in between. This way you can monitor your rabbit's digestion to ensure there was no stomach upset. If you see runny stool, smaller than normal poops, or any signs of stomach upset, including bloat, you will need to treat these issues, and refrain from giving this item to your rabbit again. Contact your veterinarian immediately.

Bunny Bits

If your rabbit has not eaten or pooped in 24 hours, see a vet immediately!

FRESH WATER

Fresh water is vital to your rabbit's health. A rabbit will drink a surprising amount of water daily. A 4-pound rabbit will drink as much in a day as a 20-pound dog. Now this may change drastically depending on the diet that you feed them. As you can imagine, a rabbit that only gets dry pellets and dry hay is going to drink a lot more than a rabbit that is getting fresh herbs, lettuces, and a variety of vegetables. If they are getting a varied diet like this, they will be getting a lot of their water intake from their foods, but multiple water sources should always be available as well.

Water should always be kept cool and fresh. A rabbit may not drink hot water, which is important to remember any time of the year, but especially during the summer. Rabbits will increase their water intake when it is hot, whether or not they are indoor or outdoor rabbits. They would rather not drink at all than drink warm or dirty water. Make sure to change their water a couple times a day, especially when it is hot. If a rabbit does not get enough water, they can develop urinary tract issues, as well as gastrointestinal issues. Their extremely fibrous diet is very dry, and it takes a lot of liquid to keep it from blocking them up.

Bottle or Bowl?

Deciding whether to use a bottle or a bowl will depend on your rabbit. When you get your rabbit, you will need to notice what they have been taught to use. Each has a pro and a con.

If you decide to use a bottle exclusively, you will be able to keep their water cleaner as they can't get stuff into it. You can also gauge how much water they are drinking daily because many glass bottles have markings on the side to measure the amount in the bottle. However, they are more difficult to keep clean on the inside. But, if you are going to be gone all day, you can rest assured that a water bottle will provide enough and your rabbit will not be doing without. You don't have to be concerned with poop, pellets, or dust getting in the bottle, which is a plus, though you should still change the water daily.

A water bowl is a more natural way to give your rabbit water. They will generally drink more water when you give it to them this way, and a bowl is much easier to pick up and wash. However, it is always possible that your rabbit may tip it over and therefore be left without water for a period of time. So always provide water in a heavy bowl to decrease the chance of tipping it over. A ceramic bowl is a good option. One perk of the bowl is that you can put some ice cubes in it to help keep it cool longer. A drawback to using a bowl is that airborne contaminants can get in, or the rabbit can get poop, pellets, or grass in it when he or she is hopping in and out or digging around.

So, my suggestion is to consider the best of both worlds. Use a bowl and a bottle, at least for a while, and see which one your rabbit prefers. They may drink from both, and you can consider that the bottle is a backup water source, which is great, but always keep them clean. A frequent wash in hot soapy water with a good rinse is imperative. Water has a tendency to cause the vessel to have a slimy buildup that needs to be washed away.

Tips and Tricks—Buy several water bottles and during the summer, fill them less than halfway up, turn them on their side, and freeze them. This way you have a solid chunk of ice inside the bottle, but not so much that the spout is blocked off. Every day, take out a freshly frozen bottle, finish filling it with water, and place for your rabbit. Always test to make sure it works before leaving. Having multiple bottles available means you can always place a fresh bottle.

PELLETS

Pellets are a controversial topic in the rabbit world, but in my opinion, they are necessary in a limited amount. It is important to purchase a high-quality pellet, containing no less than 18 percent fiber. Here is the tricky part. You need to bypass the large 20- to 50-pound bags, which are cheaper, immediately. Rabbit pellets will spoil quickly, so purchase no more than your rabbit will be able to consume in six weeks. Also, if purchasing from a pet food store or a warehouse-type store, check the expiration date on the bag

and make sure the bag isn't months old already when you purchase it. A trick we have learned is to open the bag from the opposite end. If there is any mold building up from expired or improperly stored food it will be evident from the start. Toss it, or return it to the store where you purchased it.

While some people will correctly argue that pellets aren't a natural diet for a rabbit, it is an easy and cheap way of feeding. Keeping rabbits indoors isn't natural, either. It is important to realize that good-quality pellets are packed with nutrition, and they provide micronutrients that they may not be able to get since they aren't foraging in the wild. You can easily maintain your rabbit's fresh diet most of the time, but what if for some unforeseen reason you couldn't? Your rabbit always needs the option of pellets available, and the willingness to eat them. Not to mention, you need to keep a supply of your preferred pellet on hand. It is never safe to run out completely.

It is not safe to just randomly change your rabbit's pellet, so if you and your store are out, you will be in a bind. You can switch brands if you absolutely have to, but you will need to watch for bloat, stasis, and possible diarrhea, especially if they refuse to eat them at all, which is also a risk. The ideal way to change pellets is to slowly mix them into the pellets they are currently eating. Over the course of several weeks, you start adding a few more of the new pellets to their mix, until eventually you are giving them all of the new pellet. Slowly transitioning over the course of a month is the best.

Young rabbits can begin to receive pellets as young as seven weeks old, and they should be offered alongside unlimited hay until they are at least seven months old. At this point you may need to decrease the amount of pellets; pay attention to the instructions on the bag of feed that you have. This encourages your rabbit to consume more hay, which should make up the bulk of their diet. However, you will know your rabbit best, and this can be customized based on their unique body style, activity level, and weight. This diet is suggested to help maintain a healthy weight.

HAY

Hay is a grass product that is produced during good weather months to help supply livestock with nutritious feed during cold months when nothing will grow. Since timothy hay is the primary dietary hay recommended for rabbits, we will focus on it. However, there are many other hays that are great to mix in to add variety to their diet. It is important to offer them this variety, and it will encourage them to munch more. We purchase timothy, orchard, oat, and meadow hay, and provide a nice daily mix of this.

The production of timothy hay is actually an amazing process. Generally speaking, a farmer can depend on two cuts on a field of hay per season. Depending on the current weather, they may get a third cut. It normally begins to flower in late June, and the pollen that it produces is actually a major allergen.

First-cut hay is very high in fiber and protein. Right before the first cut, protein levels can reach up to 18 percent, and can fall to as low as 4 percent by a third cutting. Some brands do not differentiate between cuttings, but you can usually tell by how eager your rabbit is to eat the hay. First-cut hay is usually coarser with thick stems because it had cooler temperatures and more access to water from spring rains. After the spring growing season, the farmer will cut the grasses and let them fall and dry out on the ground. This usually takes around four to seven days depending on the weather and temperatures. This process may be extended if the grass has to be turned to thoroughly dry out. Next, first cut is generally baled, and can be given to rabbits, but it appears most farmers use this cut for horses, and it is exported.

Second-cut hay is normally harvested in July/August. This cut may be softer, not coarse. There will be more leaves on the stems, and the stems will be much smaller. It will smell and taste sweeter, and it will be higher in calories. Also, the fiber content will generally be lower than the first cut. However, second cut is normally what you find prepackaged in the stores for rabbits and other small animals.

I cannot stress the importance of hay enough. Some rabbits love it, some rabbits ignore it, but all rabbits desperately need it. Hay is the most important part of a rabbit's diet and should make up approximately 80 percent of what they eat on a daily basis, and they also use it for bedding. It should be offered in unlimited quantities, and always be readily available. Hay is important because it supplies valuable fiber/roughage that will help them maintain a healthy balance in their intestines. This balance is essential to keep them from developing hairballs and other obstructions.

Very young rabbits are offered alfalfa hay to start out learning to eat. Alfalfa hay has a sweet flavor and entices them to eat. It is higher in protein and calcium than timothy hay, but it is lower in the fiber that they need as they get older.

When you get your new bunny, find out what type of hay they are currently being offered. If they aren't currently eating alfalfa, don't start them on it. There are several different types of hay available: alfalfa (actually a legume), timothy, orchard, meadow, oat, grass, and several companies produce a mixture. I suggest that you purchase smaller amounts of each and actually mix them. This will give your rabbit a great variety to dig through, and you may notice that your bunny will favor one over the other. Just like the pellets, don't buy hay in massive quantities. Buy bags that you can open and completely use up in just a few weeks. This way the hay maintains its freshness, color, and smell, and will entice your rabbit to eat it.

Bunny Bits

Make sure to fill your rabbit's hayrack every evening; they love to eat hay in the very early mornings.

One fun way to help encourage your rabbit to eat more hay is to use a large box or a cat litter box with high sides on it. Put a small amount of litter in the bottom, and then pile the box full of a variety of the hay. You will soon notice your rabbit hiding out in this box enjoying a snack! You can place this box in their area because they will probably get some hay on the floor around this area. A bonus of this "hay box" is that they will probably begin using it as their litter box. Rabbits love to munch on hay while they are on their litter box.

Bunny Bits

Your rabbit should eat an amount of hay that is approximately the size of their body, every day. (Not the weight, just the size.) This is why it is so important for there to be an unlimited amount available for them daily.

FRESH FOODS

Next, I give my rabbits a daily supply of fresh foods. Rabbits have approximately 17,000 taste buds, and can tell the difference in sweet, salty, sour, and bitter. How boring for them to be limited to hay and dry pellets. It isn't alway easy to vary their diet, and it can be expensive, depending on the time of year. This is one reason that I grow an herb garden especially for them. From early spring till late fall, I can supplement their diet directly from my garden.

I never give iceberg lettuce to my rabbits because it contains lactucarium, which is harmful for them in larger quantities, and it has very little nutritional value. I do give romaine and other baby lettuces that are dark green because they are higher in fiber and important nutrients. Spinach can be given sparingly, but it is very high in calcium and can create issues such as bladder sludge and stones. Cabbage is another green that I choose to avoid because it is gassy and it can create issues with bloating. Most grocery stores will have an organic premixed selection available, but green-leafed and red-leafed lettuce, Bibb lettuce, parsley, and even dandelion greens are all good choices.

Also, be aware that your rabbit is a unique personality, with her own set of taste buds. Rabbits may turn their cute little noses up at anything you offer them and flatly refuse to eat it. Don't be surprised if they refuse something you have heard other rabbits love.

Bunny Bits

There is a large list of safe greens and veggies available at the House Rabbit Society website, if you need other ideas (www.rabbit.org).

Organic or Not?

The amount of pesticides, herbicides, and fertilizers being sprayed on our food is astounding. Whether you purchase organic produce for your rabbits or not will ultimately depend on what you can afford, and what your rabbit will eat. The importance of fresh foods in their diet is undeniable. Of course, organic will be the healthiest option for

Local Organic Herbs

Rosemary

Sage

Chives

Mint

Thyme

Oregano

$2.99 ea
Small Potted

Purple Basil

Basil

them, but if you simply cannot afford to do a variety of organic options, then do the best you can. Fresh foods of any kind are better than no fresh foods at all. You will just need to pay attention to what your rabbit will and will not eat, and learn this so you are not wasting anything. Believe it or not, they will sometimes turn their cute little noses up at some items that they regularly enjoy. We generally think when this happens that they are detecting something on them that they are refusing to eat. A rabbit's sense of smell makes it especially easy for them to detect sprays and chemicals.

During the spring, summer, and fall months, you can easily grow a supply of herbs and some lettuces to supplement their diet with, at a fraction of the cost of grocery store organics.

Rabbit DIY

I choose to clean my produce this way, whether or not it is organic, just to be safe.

The easiest and best way to remove pesticides from produce is with a white vinegar/water soak. As soon as you get home from the store, while the produce is basically at room temperature, begin to fill the sink with some cool water. You will want a ratio of 1 part vinegar to 3 parts water to soak all your vegetables. Leave soaking for at least 10 minutes. Then you will need to re-rinse them again in cool water and lay them out to dry on the counter. You may need to flip them over a couple times. Once they have dried off some, put them in a clean ziplock bag with a damp napkin. You can have your rabbit drawer filled with produce that is ready for them to eat. This also saves you lots of time as you are feeding them daily. (You can buy gallon jugs of white vinegar for just a few dollars.)

Herbs

We don't know if you grow herbs, but we hope that since you are thinking about getting a rabbit or currently have a rabbit, that you will consider it! Herbs are easy to grow, and there is so much satisfaction in going out to the garden and snipping some for your rabbits—or for yourself! Not only are they healthy for you and your bunny, but they also smell divine, and most of them have lovely flowers to enjoy not only outside the home, but inside as well. Many herbs are also perennials, depending on which planting zone you are in, which is always a plus. I can be in a not-so-happy mood, and when I brush by an herb, or cut one to bring indoors, the wonderful aroma always makes me feel calm and a little bit better.

Herbs are a powerhouse of nutrition. We love to grow herbs for our rabbits, and it is such an easy little garden to maintain. Whether you can devote a section of yard to this, or a small container on a porch or windowsill, snipping fresh herbs to offer your rabbits will add so much variety and nutrition to their daily life. Fresh herbs and lettuces are also a great way to add more liquid into their diets.

Starting out, I suggest only growing and offering the culinary herbs. They are normally safe, easy to grow, and you will find yourself using them also. Plus, once your rabbit gets used to eating them, you can pick them up in a grocery store or market during the off season.

Before I tell you about the individual herbs and their possible uses for your rabbits, let me stress the importance again of calling your vet if you feel something isn't right. I use herbs as a preventative, but I do not depend on them for a cure. If your rabbit is sick with any issues, whether it is stomach related or respiratory, please go to the vet immediately. You can offer these herbs as a supplement to the medication that the vet will give you. Rabbits get sick very fast, and waiting even an extra day may be too long. You can offer herbs daily all during the year to your healthy rabbit as a preventative to help ward off any unwanted illnesses.

We have a raised-bed garden that is completely fenced in. We don't have to worry about the possibility of wild rabbits or other animals getting in and contaminating the area. Anytime you take your rabbit outside to let them hop and play, there is always the possibility of it coming in contact with an area where a wild rabbit, or any animal for that matter, has been. Other animals can carry diseases that your rabbit has not been exposed to, or that your rabbit could catch, such as *E. cuniculi* and *Pasteurella*. If budget and space allow, consider setting up a protected little garden area for your bunny to play in, or at least a place where it can go dig and nibble on fresh edibles. A few large pots on the deck filled with some of their favorites would be another option. You could also grow

some potted herbs in the laundry room or on a kitchen windowsill. Sugar loves any time he gets to spend in the garden. It is always fun watching him enjoy nature and spending some time outdoors, closely supervised, of course! We generally devote one small section of my garden to lettuces and kale. We grow parsley, sage, thyme, lemon balm, chamomile, dill, and cilantro. We also grow echinacea to feed both fresh and dry to mix into their hay mixtures. Any of these herbs can be dried out and used as treats later.

Even if you decide not to grow your own herbs, most of them are easily accessible fresh from the market. Always give them a quick rinse before offering them to your rabbit.

Beginner Herb Garden

Here are a few suggestions of herbs to grow for a beginner herb garden. These herbs are very easy to grow and very forgiving. This is a great place to start, and you can add as much as you want. You will find a few more suggestions on page 67 if you want a larger garden. If you don't see something listed, do a bit of research prior to giving it to your rabbit, just to make sure it is safe for them.

Parsley is easy to grow, and rabbits seem to really love it. Only feed a small amount to them because parsley is high in calcium and large amounts can contribute to bladder sludge and other issues. However, in small amounts parsley is an excellent treat, and it is believed by many that it can be used to treat a myriad of urinary tract issues. Parsley has a fresh grassy smell and rabbits can eat both the leaves and stems.

Sage is also easy to grow, and rabbits enjoy nibbling on fresh or dried leaves. Small amounts (meaning two to three leaves with stems) is plenty for a small serving, and sage is an excellent stimulant for the digestive tract.

Thyme is an excellent treat to give when your bunny seems to have a digestive issue going on, as it is very soothing. You can feed the leaves and stems (before the stems become woody later in the summer).

Lemon Balm is a very calming herb, and one I use constantly to try to soothe stomach issues, such as bloat, before it starts or gets bad. It is also believed to have antiviral and antibacterial properties.

Chamomile is a wonderful herb to grow, but most of the time I purchase this dried. It is very hard to grow enough chamomile to offer as treats. However, I use it constantly to help reduce stress and lessen pain. I have also made a mild tea out of just chamomile before and offered that to entice a rabbit to drink. The tea is also an excellent eyewash if your rabbit has a weepy eye, or something has irritated it. I just take dried chamomile, put it in a sterile tea bag, brew it in warm water, wring out the bag a little, and carefully lay the warm (not hot) bag against their eye.

Dill is a nice, fresh snack for your rabbit. Dill helps support intestinal motility and cecotrope pellet formation. It has fewer calories and fat than fruit and non-leafy veggies, and it is bursting with vitamins, minerals, fiber, and antioxidants. Your rabbit may not find dill to be tasty, but Sugar loves it! He gets so excited when I bring him fresh herbs from the garden! Just experiment by offering a variety of different herbs to find out which ones are pleasing to your rabbit's palate.

Echinacea is another powerful herb that you will want to keep around. Whether fresh or dried, it is a great way to boost their immune systems. You can give the leaves or stems mixed in with their hay. This can be given as a preventative or if your rabbit is currently sick.

Cilantro is an herb known for its antioxidant properties. It promotes a healthy digestive system, and helps prevent urinary issues. The essential oils and rich aroma of freshly snipped cilantro are excellent for getting the digestive juices going, and have antimicrobial agents that help fight off common infections.

Always research any herbs or other foods that you are considering giving your rabbit! There are a myriad of foods that can be toxic. Better safe than sorry. Keeping your herbs cut back and snipping them daily for your rabbit will help them to continue growing through the season. Most herbs are drought tolerant, which is another plus for water conservation, and most of them can be grown in a sunny location. Always keep your rabbit's feeding station clean and remove any old, uneaten herbs left more than a day.

Bunny Bits

When planting an herb garden, plant twice what you think you will need! Your rabbits will go through it very fast, and by midsummer you will be trying to find more to plant!

Bunny Bits

Another great treat you can add to your garden or a pot on your porch is carrots. Of course carrots are a favorite treat for most rabbits, but this time, you are growing them for the tops. The tops are another green that you can add to your rabbit's diet. Once they have a nice top, you can cut them off, and they will regrow.

The Rabbit Drawer

From our research, and observing what a wild rabbit prefers, a rabbit requires a combination of foods to keep its digestive tract working at peak performance. So, in addition to their hay and pellets, I have a rabbit drawer in my refrigerator. This drawer contains only things that are for my rabbits, and are safe for them to eat. It is important to realize that not everything is safe for your rabbit to eat. Having a separate location for their items is a safe way anyone can actually feed them if something happens and I am not at home, and fresh is always best for them, just like it is for us.

Anything in the rabbit drawer should be safe for them and ready to eat. When I come home from shopping, I will fill the sink with all the produce I purchased for them. I will then rinse everything off in a vinegar/water mixture and then soak it for a while. After a final rinse off, I lay everything out on the towel to let it dry off for a while. Then I bag it up individually in plastic bags. If you want, you can even get small ziplock bags and prepare meal bags for them that are ready to go! I don't always keep bags ready. In the spring and summer when the garden is plentiful, I love to go pick their daily salad fresh.

Here are a few suggestions for making your bags up; however, this all depends on your rabbits' tastes and what they can tolerate. You will have to slowly discover your rabbit's likes and dislikes. Adding a variety and switching it up is a great way to keep your rabbit excited for their mealtimes. I have one rabbit that doesn't like any fruit at all except apples. So, her bags only contain an apple slice for her fruit. Also, this is only for fresh food—unlimited hay and a correct amount of pellets should be available at all times. There should never be a time that there is no food available for them in their area.

Bunny Bits

Always make sure to remove all apple seeds before giving any to your rabbit! Apple seeds contain cyanide, which can poison your rabbit.

Bunny Bits

If you are away and have someone rabbit sitting for you, these pre-made bags are a great way to have meals ready to go for them. This will ensure the fact that your rabbit receives the proper amount and variety of food each day that you aren't there. They become very dependent on the diet they are used to, and if they go for days without their normal supply of fresh foods that are high in fiber, it can aggravate their sensitive digestive tract, which has become used to a certain amount.

Bag #1
a few pieces of kale
three sprigs of thyme
two pieces of dill
one slice bell pepper
one small carrot

Bag #2
romaine leaf
two stems of parsley
organic spring mix
cauliflower leaf
one section of apple

Bag #3
a few pieces of kale
one piece asparagus
carrot tops
organic spring mix
one inch banana

Bag #4
romaine leaf
dandelion greens
celery stalk
one cherry tomato

Bag #5
kale
two sage leaves
fresh grass (handful)
two blueberries

Bag #6
romaine leaf
basil
one broccoli stalk
dried papaya slice

Treats are a great way to train your rabbit and strengthen your bond with them. They have an incredible sweet tooth, so you can teach them quickly to do things like return to their cage if they recognize the treat in your hand. One thing to realize is that anything that is offered and hand-fed to them is a treat. My bunnies think it is special when I hand-feed them choice pieces of their hay. Yes, they would prefer raisins, and I do give them special sweet treats, but it isn't always necessary. Just make sure you never give them too much. Too much of a good thing is really bad for your rabbit's health.

You can also make some healthy treats with recipes listed later in the book. A few reputable companies are making very healthy treats also. Oxbow offers treats and supplements that are specifically formulated to treat issues like digestive, skin, joint, and urinary tract problems, among other ailments.

Bunny Bits

Dried papaya (without added sugar) is a great treat for your rabbit, and it is believed to help with digestion. You can also use it to help train your rabbit to return to its cage on command! Just make sure there is no added sugar.

Suggested Greens, Herbs, Vegetables, and Fruits
(Items with an * should be offered or given infrequently, and in small quantities.)

GREENS	HERBS	VEGETABLES	FRUITS*
Arugula	Basil	Asparagus	Apple (no seeds, stems)*
Borage Leaves	Calendula	Bell Pepper	Apricot*
Bok Choy	Chamomile	Broccoli*	Banana*
Cabbage (Dk Green)*	Cilantro	Brussel Sprouts*	Berries (uncooked)*
Carrot Tops	Dill	Carrots	Cherries (no pits)*
Chicory	Echinacea	Cauliflower Leaves	Currants*
Dandelion Greens	Fennel	Celery	Grapes*
Dill	Lavender*	Cucumber	Kiwi*
Endive	Lemon Balm	Eggplant (no leaves)	Mango*
Escarole	Mint	Pumpkin	Melons*
Frisée Lettuce	Oregano	Squash	Nectarine (no peel)*
Fennel	Parsley*	Zucchini	Papaya*
Kale	Rosemary		Peach*
Mâche	Sage		Pear*
Mustard Greens	Thyme		Plum*
Romaine Lettuce			Pineapple*
Raspberry Leaves			Raisins*
Radicchio			Tomato (no leaves, stems)
Spinach*			
Spring Greens			
Turnip Greens			
Watercress			
Wheatgrass			

As a quick reference, here is a list of things that are **toxic** for your rabbit to consume. This isn't a complete list, and you should always do a little research before feeding rabbits anything new.

TOXIC FOODS	BAD VEGETABLES	BAD FLOWERS	BAD WEEDS/HERBS
Avocado	Cauliflower Florets	Aloe	Chives
Bread	Chard	Amaryllis	
Cookies/Crackers/Cereals	Iceberg Lettuce	Buttercup	
Chocolate		Jasmine	
Meat		Lily of the Valley	
Onion		Milkweed	
		Eucalyptus	
Potatoes			
Rhubarb			
Fruit Pits/Seeds of any kind			

Herb Bricks

Parsley is especially good for your rabbit's health. It is believed to help improve urinary tract health, helps treat inflammation, benefits fertility, and improves blood count. Parsley leaves and roots can also be used to treat constipation.

If you don't have parsley on hand, you could replace it with any rabbit-safe herb that you currently have an abundance of in your garden. You can add a tablespoon of Oxbow Critical Care if you have that on hand also.

Another added benefit of these bricks is the fact that they are hard and brittle. This is very good for your bunny's teeth! Just make sure to give them as treats, not in place of their usual diet. I usually give one per day per bun!

- 1 cup rolled oats
- ½ cup rabbit pellets
- 2 medium-sized bunches of parsley
- ½ carrot
- ½ banana
- ¼ cup water

Grind your oats and pellets in a food processor until finely ground.

If you have a large enough food processor, add the parsley, carrot, banana, and water, and grind again. If not, grind ingredients separately, and then mix together in a bowl.

Bottom right photograph shows the consistency the mixture should make. If you need more moisture, you can add more water, or more banana. Cover with wax paper and flatten the mixture to about ¼ inch. Score the mixture with a knife, making small squares.

Lift wax paper and lay it on a cookie sheet. Bake at 350°F for approximately 35 minutes. Do not let it get too brown. Next, simply turn off oven and let it completely cool with the bricks still in there. You are attempting to completely dry them out. Any moisture means that they could grow mold over time. If you still feel like they are moist, remove from oven, reheat oven to 350°F, and turn it off again, and set them in for the cool down.

Another option would be to place them in a dehydrator at this point, if you have one.

Once they are completely dried out and brittle, simply snap them apart.

Store them in a jar or ziplock bag, or anything airtight.

If you're making more than one batch, store the bulk of them in the freezer. That way, you can have herb treats for your buns all winter long!

Dehydrated Carrot Chips

My bunnies love these, but carrots are high in sugar, and dehydrated carrots have an even higher concentration of sugar. So, these are special treats for very good bunnies. I usually give one per day, so one batch will last for a long time.

- 1 very large, fat, organic carrot
- Dehydrator
- Sharp knife

Scrub the carrot and slice it very thin.
Arrange slices on dehydrator in single layers, not touching.
Turn dehydrator on high for about 12 hours, or until the carrot slices are dry and crispy.

Summertime Cool Treat

During the hot summer, your rabbit will enjoy a cool treat, and these are very easy!

Take an ice cube tray, and fill it up with chopped herbs and fruits. Then, add water to fill the container and freeze. Once frozen, pop them out and place in a ziplock and take one out as a treat on hot days. I place them in a small dish so that as they melt, it provides cool water, as well as tasty cold fruit to help cool them down.

OVERWEIGHT/UNDERWEIGHT

As you can tell, what you feed your rabbit is going to occupy a great deal of your thoughts. Watching their weight will also become a concern. Their ideal weight will fluctuate based on breed and body type. Your vet can help you determine what their ideal weight is, and help you adjust their diet if there is a need. Being overweight can contribute and worsen many health issues, and being underweight can be a first signal of many illnesses. So, recognizing their ideal weight as well as their current weight is a great start to maintaining their health. Many vets recommend weighing your rabbits at home, as this will help you monitor their general health.

IS MY RABBIT EATING ITS OWN POOP?

Yes! One thing you will notice early on is that your rabbit will eat his own poop! Rabbits produce two types of droppings. The first is the normal fecal pellet that resembles a small, brown pea. When you pick it up, if their diet and digestion is in good shape, it will be dry, and will crush easily with pressure. The other dropping is called a cecal pellet, or a cecotrope. These small droppings are actually the fermented remnants of plant material that have been digested and redirected into their "cecum." The cecum chamber is attached to the large and small intestine, and it contains a specific amount of bacteria and fungus that is needed to take this plant food material, ferment it, and create a nutrient-rich pellet that is full of vitamins that the rabbit would not be able to get any other way.

Rabbits normally produce cecal pellets every day, and they resemble a small bunch of grapes. The number of cecal pellets they produce a day will depend on their diet. They are usually sticky, slightly wet, and smelly. You may not ever notice this because a healthy rabbit usually ingests this pellet as it is coming out. You may see your rabbit sitting, appearing to bathe their lower abdomen, when they are actually eating a freshly digested pellet.

Bunny Bits

Notice the normal size, shape, texture, and amount of your rabbit's poop. Recognizing what their healthy poop looks like will help you identify signs of illness early on. One of the first signs of GI stasis is a change in poop size, shape, and amount.

Sleeping Habits

Most people think that rabbits are nocturnal (sleeping during the day and staying awake at night), but they are actually crepuscular. This means that they are most active during the twilight hours and at dawn. In the wild, this would be the safest time for them to leave the burrow and forage for food. Your indoor rabbit will have these same habits to a certain extent, but they will adapt to the schedule they see around them. A rabbit normally sleeps between 6 and 10 hours during a 24-hour period. You will notice them actively eating, grooming, playing, exploring, and relieving themselves, and napping throughout the day. You will often see them in what I call a "loaf" or a "lump" napping with their eyes partially open. Sometimes you will see them in a complete bunny flop (often looking like they are dead) on their side in absolute total abandon. However, it doesn't matter how asleep you think they are, they are able to come to complete attention at the slightest sound or movement in their environment. Sugar likes to awaken me at 5 a.m. every day

for his breakfast. Don't worry, I am an early riser. You and your bunny will work out a schedule that will suit you both, but don't be surprised if you get a thump if you oversleep or don't get them their bunny breakfast soon enough! By around 10 a.m., Sugar is ready for his first nap of the day.

Rabbit Care

Keeping Their Environment Clean

Whether you have decided to keep your rabbit indoors or outdoors, the importance of keeping everything clean cannot be stressed enough. Simple daily/weekly cleaning will help ward off so many sickness and stresses, and it will make for a much happier bunny! Remember how strong their sense of smell is? Your cleaner of choice should always be white vinegar. It is safe for the rabbits, does a great job cleaning, leaves no smell, and it neutralizes urine odor. Cleaning all aspects of their environment at least once a week is the best way to start, but some things may require daily touch-ups.

SAFE CLEANING

The first thing I want to discuss about cleaning is how toxic it can be. When getting ready to clean or sterilize your rabbit's cage, housing, or your home, it is important to remember that anything you use can be ingested or inhaled by your rabbit. Many things are completely toxic, but the thing that is the safest for you to use is also the cheapest. I can purchase a gallon of plain white vinegar for less than four dollars, and it will last a long time. Just get a spray bottle and mix it 50/50 with water for basic cleaning. You can also use it straight if you need to sterilize something. You can use the vinegar to clean the water bottles, food dishes, cage floor and walls, and even in the laundry when washing bunny blankets or clothes.

Things that you should never use around your rabbit are any type of harsh or synthetic chemicals. Keep in mind that anything that they walk on, they will lick off their feet, and it has the potential of getting in their eyes. Anything that is in the air, they will breathe, and neither their respiratory systems nor their GI tracts will be able to stand up to strong chemicals. Rabbits are obligate nasal breathers, making them very sensitive to anything that is in the air. This is important to remember when purchasing products to mop the floor, clean the carpet, or even spray into the air.

The number one thing you must **not** ever use in any way is baking soda. While baking soda is normally considered to be an all-natural, safe cleaning alternative, it is deadly to rabbits. Many small-animal bedding products and small-animal litters may contain baking soda, so it is vital that you read the ingredient list prior to using anything around your rabbit. Baking soda that your rabbit ingests will cause a massive buildup of gas in its stomach. Your rabbit has no way to expel this gas, as they are not able to burp, and it will cause a horrible case of bloat that will be very painful and most likely fatal. If you are using baking soda products to clean your carpets, it can severely damage their respiratory system, even causing pneumonia.

Diffusers and candles are extremely popular, and we do use them sometimes. We are very cautious about the area we use them in, and where the rabbits are at the moment. Don't diffuse oils, burn candles, or even use the aerosol sprays in small enclosed areas. I know it is tempting, but these items are airborne, and they will land on their fur. In turn, once on their fur, they will be ingested. We are not saying you can't use them at all, but I wouldn't run them constantly, and I would make sure it was in a well-ventilated area. Also, choosing all-natural products will keep many of the additives that are the most dangerous out of the air and off of your rabbit.

If you use blankets for your rabbits, make sure that you wash them in detergent that is all-natural, free and clear of any dyes and scents, and, of course, contains no baking soda. There are several options on the market. You can also mix a cup of white vinegar into your washing machine to help neutralize any smells or ammonia that are in their blankets.

If your rabbit has come into contact with or, especially, ingests any chemicals or baking soda, take them directly to the vet. It is an emergency!

LITTER BOX CLEANING

If a litter box isn't kept clean, it will create all sorts of issues for your rabbits. A litter box should be changed based on your rabbit's daily use. Some rabbits can go for a week between changes depending on the type of box you have, but some rabbits will insist on a daily cleaning. It will be their preference. However, if you have a very clean rabbit, and you don't do daily changes, you may discover they will lose their litter box training, preferring a clean place. A dirty litter box may also mean that

their poop will stick to them when they enter, which causes their fur to be dirty, and they will be accidentally dropping them outside their litter area. Finally, a dirty litter box will also attract flies, which will greatly increase the risk of warbles. Most people think that rabbit litter stinks, but this is only if you don't do a decent job cleaning. Whether you have a box that has a grate on it, which separates the poop from them, or you have a large box full of hay, you will always need to clean it at least once a week; twice is better. Once dumped, you can use white vinegar to scrub it out. Once dried, refill with fresh litter and replace the box quickly. Rabbits get very annoyed when their litter box is "closed for cleaning."

Water Bottles and Bowls

Fresh, clean water is essential to your rabbit's health. A rabbit may drink up to 10 percent of its body weight daily. However, once the water isn't fresh, rabbits don't drink as much as they need. They will always prefer cool water from a clean container. Once again, cleaning with white vinegar will be the safest way. Change the water and clean the container daily, making sure to rinse it well to remove all vinegar. During the summer, you will need to change the water even more frequently, especially if you have an outdoor rabbit.

Food Bowls and Hayracks

Food bowls also need to be kept clean. Fresh food bowls can get grimy quickly, and even bowls to hold pellets will get dusty. Any unclean area that contains food will quickly attract bugs. Hayracks aren't difficult to keep clean. I add fresh hay daily, and toss everything that wasn't eaten once a week. Hay, even dust-free hay, tends to be dusty. This can cause eye issues and sneezing. So, maintaining a clean environment will prove beneficial not only for your home, but also for their health.

Bedding

Even though my outdoor rabbit has straw bedding and my indoor rabbit has a blanket, I choose to clean out both completely once a week. I will change either at any point if I notice them soiled. Dirty bedding will once again attract flies, fleas, and ants. If you use pine shavings or straw outdoors, keeping it fresh will be the best way to help keep your rabbit healthy and free from mites, fleas, parasites, warbles, and fly strike.

Indoors isn't much different; a dirty blanket or litter box will attract flies. Also, make sure that their blankets don't have any holes in them. Rabbits can get tangled up in the holes and panic. Worse yet, they can get the hole around their neck if they are digging in their blanket. This could actually choke them.

FOOD STORAGE

You should never purchase more than six weeks' worth of rabbit pellets or hay at a time. Any older and it will become old, and if there is any moisture in your environment, it can begin to mold. This is terribly dangerous, and moldy food can cause death quickly. All food should be examined and checked for freshness each week. Any food older than six weeks should be tossed out. At least once every six weeks, all containers should be completely cleaned out with vinegar, rinsed, and allowed to completely dry inside. Once you are sure it is completely dry, you can replace the food and start over.

Keeping the Indoor Environment Safe

So, you have decided to keep your rabbit indoors. Wonderful! Something you may not realize is that even though you are keeping your rabbit indoors, there are things you need to do to keep him or her safe. First, I want to discuss two things that are dear to my heart. I have experienced these issues firsthand, and two of them involve the kitchen. Did you know that it is harmful for indoor pets to be in the kitchen while you are cleaning your oven? Especially using the factory setting on the oven for self-cleaning. I lost six precious chicks that I was housing in a kitchen nook doing this. Even opening a window during this process doesn't help. So, save yourself a lot of heartache and don't clean the oven with your bunny (or other pets) in the house.

The other instance occurred one day while I was frying something on the stove and didn't have the exhaust fan on in the kitchen. Before I knew it, the room was full of smoke! Thankfully, I thought about my bunny, Sugar, and looked over to make sure he was all right. He was in a panic, darting back and forth in the adjacent family area! I immediately opened the windows and doors and turned on the exhaust fan, scooped him up, and carried him outside for some fresh air. It didn't even dawn on me that the smoke would harm him, especially since nothing was burning at the moment. I felt so bad. So, always make sure you have adequate ventilation when cooking!

Do you have stairs? We find using a gate at the bottom of the stairs is the safest thing to do. A rabbit will not have any problem getting to the top, but when they decide to come

down, they will have a difficult time and tumble down partway, or even all the way. We keep our female rabbits upstairs, and several times Sugar has made his way to the top to see them. He has fallen down several stairs trying to get back down. Stairs are slippery if you don't have carpet on them, but even if you do, they can still have problems getting down. A rabbit hops or jumps straight out. They can't make that adjustment to take one step at a time on their descent, like a dog or cat. So, keep that in mind, not only where stairs are concerned, but also with stools or chairs. If they jump up on a stool, they need ample room to jump out, without fear of jumping on another object while descending and possibly injuring their skulls, or even breaking their teeth.

Bunny Bits

Many rabbits can manage stairs fine if they are carpeted. They just have to learn how; it is wood or slick surfaces that generally create the problem.

Electrical cords can be a hazard for any pet that likes to chew, be it a puppy, kitten, or rabbit. So, always be on the lookout for any frayed cords. Cord chewing may result in a fire and could cause electrocution. Anything that your rabbit could get tangled up in should be considered. They have a tendency to panic when they get their feet or legs caught and this could result in broken bones!

One hazard that you may not think of initially is houseplants. Many houseplants will be toxic for your rabbit, and believe me, they will be attracted to them! They love the smell of fresh things and dirt. If you have a potted plant within their reach, they will help themselves to it. So, just to be safe, move all plants into an area of the house that they are not allowed into. Even if you set a plant up high, leaves or petals that fall off look like a tasty treat, but they may be deadly.

A lot of indoor safety considerations are just common sense. Sugar loves to do what we call "round and round." Yes, he knows what it means! It is so cute. Now, it always happens in the middle of the family room floor. All we have to say is "Do you want to go round and round?" and he's off! I will walk in a fast motion around the center of the room, and he begins running with me. Although, not necessarily always in the same direction that I am going in. He will run back and forth in a straight line, or in circles counterclockwise. Sometimes he will even make a new trail around the ottoman until he stops short in complete exhaustion. So, if you and your bunny decided to go "round and round," please make sure he or she doesn't get tangled in your feet!

As we said, keeping a bunny safe indoors is mostly common sense. So, just take a deep breath, relax, and enjoy your special little bundle of fur!

Litter Box Training

Litter box training your rabbit is an extremely important part of maintaining a clean environment for your rabbit. Rabbits are naturally clean animals and prefer to keep their living areas clean also. We are going to rethink the concept of training. Your rabbit will litter box train you (more on this below). When you first bring your bunny home, it will be important to keep them in their new living area, temporarily, while you are being trained. They can get out to play, of course, and explore some, but should be returned to their area to ensure that they learn this is where they are to eat, drink, and go to the litter box. If you have given them appropriate housing, then they will have a nice area to be active in while they are training you.

Litter Box Tips:

1. A large cat litter box will make an excellent place for your new rabbit. Make sure that you only use rabbit-appropriate litter. Clay litters are dangerous for rabbits. You can locate what is called small animal litter, or even litter made from compressed paper or wheat straw. (Also, avoid any cedar- or pine-based beddings.)

2. Put several large handfuls of fresh hay in the litter box, or put a hayrack over top. Rabbits love to eat their hay while they are on the litter box. Hanging it close will encourage this behavior. Always keep a fresh, unlimited supply of hay available for them.

3. Make sure to gather up all poops you find scattered around their new home and place them and your bunny in the litter box.

4. Next, you will possibly need to find a spot where they have urinated in their area, and absorb it with a paper towel. Place this in the litter box and cover it with a small amount of litter and hay.

5. You will soon notice that your rabbit is either choosing to use their new litter box, or they are creating a pile in another corner of their choosing. Here  is a very important part of your training. Instead of attempting to train your rabbit to change their location, you will move the litter box to the spot they have chosen. You may have to do this a couple times, but eventually you will agree on the spot your rabbit chooses. Make sure to use a calm, sweet voice to give positive reinforcement anytime you see them go to the litter box on their own.

6. Once they are consistently using the litter box, or at least getting close by remembering to return to their area, you can enlarge their free-roam area. If they forget their litter box habits, then return them and their droppings to their area. Place both in the litter box, and give them affirming words in a calm voice.

7. Never discipline a rabbit by any means other than calmly and gently returning them to their area for a little while. They will not understand, and a loud voice can terrify them. This can create a block in your relationship that you may not overcome. Any physical discipline can severely hurt them as their skeletal structure is extremely fragile.

8. Finally, if your rabbit is having difficulty with litter training, be patient. Sometimes young rabbits don't figure this out quickly. Once they are spayed/neutered it will become much easier and more natural for them.

Grooming

Grooming is an important part of your rabbit's health. The amount of grooming required and the frequency will depend on the breed of the rabbit you choose, their personal cleaning habits, and the environment that they are kept in. Rabbits are continually grooming themselves, and they prefer to be extremely clean. Rabbits view grooming as a social activity; they groom each other in wild colonies. So, take time while grooming and your rabbit may learn to enjoy the interaction.

Basic rabbit hygiene care will include brushing, nail trims, and checking the ears. If you have had a bunny who has had a round of diarrhea, you may have to do a "butt bath," and spot clean.

Bunny Bits

Please be careful if you decide to take your rabbits somewhere to have them groomed. Grooming a rabbit is very different from a dog or a cat, and most places have never done it before. If you need help, try finding someone locally who has a rabbit and can help you. We have heard too many stories of rabbits being seriously injured by someone grooming them who had no experience.

BRUSHING

Let's discuss brushing first. If you have a short-haired breed rabbit, you may only need to brush them once a week. Long-haired breeds such as Lionheads and Angoras may require daily brushing to maintain their coats and keep them from ingesting too much fur. The only exception is during their molts. A molt is when your rabbit sheds their heavy coat, several times a year. Rabbits have a light molt about once every three months. At the end of winter and the end of summer, they will experience a very heavy molt. Any molt is a dangerous time for them because all fur ingested has the potential to slow down their intestinal tract, which can lead to some bad problems. During a heavy molt, you must plan to brush them daily. Molts generally last between two and six weeks, depending on the individual rabbit and their breed.

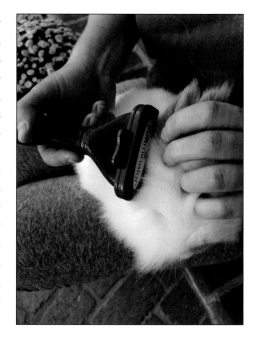

A fine-tooth brush, such as these shown, are a good choice for grooming. They seem to really be able to grab a lot of fur, and work through mats easily. While you are brushing them, make sure to be careful about the pressure you use. Their skin is very delicate, and you don't want to risk wounding them.

There is a bit more about grooming and trimming fur starting on page 132.

Bunny Bits

If you take a wet washcloth and wipe your bunny slightly—not soaking, just damp—then the fur will not fly so much as you brush them.

THE BATH

Rabbits should never be given a bath. Let me repeat. Do *not* bathe your rabbit. No matter how cute that video you saw on the Internet was, or just how sure that person on the forum sounded, bathing is not natural for rabbits. They are self-groomers. Normally, they will be able to clean anything that they need cleaned. Rabbits who are submerged in water can easily and quickly go into shock, which can be deadly. It is also dangerous for them to get water in their ears. This can easily cause ear infections, fungal infections, and so much more that can be so easily avoided.

The only exception to the "no bath" rule is if your rabbit has a case of "poopy butt," or had an accident. The "butt bath" is acceptable, but you must be careful and support their backs in case they get scared. When you give a butt bath, you should probably pick a sink that is easy for you to access running water, without having to bend over. Your focus should be on protecting your rabbit and not allowing them to kick and possibly hurt themselves. You should use warm water, and simply rinse their hindquarters off. The bulk will simply wash off. We try not to use soap if we can help it, but you can usually find a safe, mild soap that is made specifically for rabbits. Most places will tell you that a baby shampoo can be used; if we are purchasing that we search for a unscented, fragrance free, and dye free. If you have access to an organic grocery store, this will be better. You can locate a fully organic mild baby shampoo that has very few ingredients, while still being fragrance and dye free.

Once we finish, we use a soft washcloth and carefully use it to soak up the water that has collected in their fur. Rabbit fur, once wet, holds the water extremely well. You can usually towel dry the majority of the water. However, if your rabbit isn't scared, you can use a blow dryer if it has a low speed, and you can control the temperature of the heat. It would be very easy to burn their delicate skin, so turn the heat down as much as possible. It is very important to get them dry, especially around the folds of their skin. If moisture remains, it can cause fungal infections to start.

Diarrhea should be washed off immediately, as it can cause burning, irritation, bacteria, and even impaction if it is left on them for any length of time. Urine can cause the same burning, irritation, urinary tract issues, and scalding on the skin. This scalding is extremely painful, and it happens when they have urine against their skin, whether it is their fur holding it there, wet bedding, or a puddle on the floor of their cage. It must be completely cleaned off every time it happens to keep their skin healthy. Either of these issues should warrant an immediate phone call to the vet.

Any other dirty areas can easily be spot cleaned. Using an all-natural baby wipe is a great alternative to a bath, and they usually have soothing ingredients— just make sure to check the ingredients and make sure there is no baking soda in them.

Hopefully, the butt baths will be few and far between, but if you have to do them often, you may also want to have a salve to soothe dry skin that may develop. You have several options, but always lean toward the natural options. Sometimes, simply using coconut oil will be plenty, however, if you need something a little stronger, try A+D ointment without zinc. You can find this in the baby aisle of your drugstore. If you decide to use a baby powder to help absorb moisture, make sure it is cornstarch based, and contains *no* talcum.

Bunny Bits

Make sure to talk softly and be calm when giving a bath. They will pick up on your anxiety and become anxious themselves.

TRIMMING NAILS

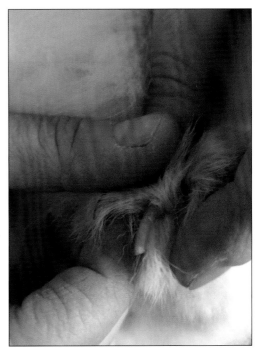

Nail trimming is a very intimidating part of the grooming process. Nails that are too long will hurt their feet, change the way they walk, and can even cause sore hock. If you are too nervous to trim your rabbit's nails, we would suggest taking them by the vet's office. Usually, they can do it quickly, and will be willing to teach you how. It may be challenging the first few times you do this because you and your rabbit are learning. They are scared that you will hurt them. The trick is to be calm, and pay close attention to what you are doing. If your rabbit is calm enough, you can place them on your lap, and hold one foot at a time, putting slight pressure on each toe until the nail is visible. When you can see the nail, you will need to look for the "quick." The goal is to never cut into this area. It will cause pain and bleeding. It is very easy to see the pink quick on a white rabbit with white nails. Darker colored rabbits will have black nails, making it more difficult to see.

- Rabbits have five toes on their front paws and four toes on their back paws.
- Make sure to use pet nail clippers that are sharp and good quality.
- Have styptic powder or cornstarch on hand to help stop the bleeding if you accidentally cut the quick of the nail.

- A small flashlight can help you see the quick if you are having trouble.
- You can also use a washcloth to dampen the fur on their feet prior to trimming their nails. This will help you be able to see the nail without the fur in the way.
- If your rabbit doesn't want to sit on your lap for this, you may need to get a blanket and do the famous "bunny burrito."
- If you and your rabbit are nervous, head to the vet. They can do it easily with less stress for you both.

SCENT GLANDS

Scent glands are one of the more disgusting parts of rabbit grooming. If you notice your rabbit all of a sudden has a very odd smell that is reminiscent of onions or strong coffee, or a combination of both, then their glands are needing a touch-up. Some rabbits do a great job of keeping themselves clean, but others need help. Unfortunately for me, all my rabbits need their glands cleaned about once every six to eight weeks or so. It is best to check them when you are trimming their nails, so you can get it all done at once. I would suggest, at least the first time, ask your vet if they can show you how to do this. This is not something you normally have to do very often.

All rabbits have four scent glands, one located on each side of their chin, and one located on each side of their anus. These scent glands are used to distribute your rabbit's unique scent signature onto their territory. This is one time when you will need to carefully hold them on their back, or in a sitting position, depending on how calm your rabbit is. You can even carefully wrap them in a towel, a.k.a., the bunny burrito. Always be careful doing this; we prefer sitting on the floor so if you lose control of them they can't fall far. Be careful to support their back in its natural curved position.

You can see in the photo how you can gently locate the glands by pressing on these areas. You will notice a waxy buildup that may appear light yellow and have a chalky texture, or a dark brown, waxy plug. We generally have a few cotton swabs, a paper towel, and some warm water on hand. It takes some practice, but it gets easier each time you do it. Just wet the cotton swab with the warm water and gently work out anything in this area. Notice the extremely thin skin in this area. Apply only very gentle pressure, and if it doesn't want to come out, try putting the warm water on a cloth and holding it against the clog to help loosen it. You never want to actually dig at this to remove it, as you can easily tear the gland. Fair warning: It will be smelly.

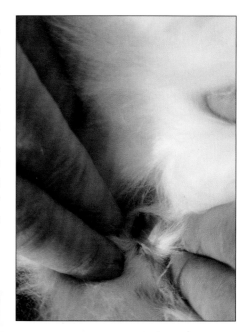

Bunny Bits

Scent gland cleaning should not be part of your weekly grooming. It should only be done if needed.

Chewing

Rabbits love to chew. It is perfectly natural. This innate desire can be frustrating for a first-time rabbit owner, but chewing is essential to keep rabbits' teeth worn down, and it is also a way to stimulate them, and help keep them from becoming bored. There are many options for chew toys that can be purchased or homemade; the more you can offer in different varieties, the better the chances that they won't decide to chew on your favorite chair. However, this is always a possibility, so for a while it is better to "bunny proof" your living areas, until your rabbit has reached adulthood. By then, you

should be able to tell if they will be content with chewing their toys, or if they will help themselves to your baseboards.

Rabbit DIY Project—Baseboard Picket Fence

It is always better to protect your baseboards, doors, and door frames before any damage happens. This is a cute little project that you can easily do to protect your framework until you know if your rabbit will want to chew these items. If you pick up a few pieces of wood and have a saw, you can make something sweet and functional that won't be an eyesore in your home.

These fence pieces are made for thin ¼" wood. By attaching some double-back strips, you can easily stick these to the baseboards around the rooms you keep your rabbit in. This simple trick not only protects the baseboards, but it also gives you a place you can tuck cords into so they are out of the reach of your rabbits.

DETERRENTS

One easy and free deterrent for chewing is an empty toilet or paper towel holder. However, it is important to make sure that any paper products are free of dyes or possible chemicals. Cardboard boxes are another alternative. There are some cute cardboard "castles" or "houses" on the Internet, or you could make your own. This would be a good project for a young child to do for their bunny. We have one piece of rolled up newspaper that Sugar likes to chew on occasionally. It barely sticks out of the magazine rack, and I see him go to it now and then. He has made quite a "dent" in it! Do you have any old natural baskets around that aren't being used for anything? Any basket should be free of shellac and paint. Below is a list of woods that a rabbit can safely chew. (Be sure the wood is chemical, paint, and pesticide free, of course!) Please notice that cedar is not on the list. Cedar is **never** suggested for use with rabbits. Please make sure to not use cedar shavings, cedar houses, or give them cedar to chew.

Safe Wood Suggestions

Apple	Birch	Blackberry
Fir	Hazel	Hawthorne
Maple	Pear	Raspberry
Spruce	Willow	Pine

Rabbit Toys

PLAYTIME—TOYS AND MENTAL STIMULATION

Rabbits are very smart. They love to play. They love toys, and they even like to make up their own games. There are some great toys available that you can purchase for your rabbit, or you can take a little time and make some. Either way, make sure to keep several available. They love to toss things, throw things, carry things, dig things, hide in things, and, of course, chew on things. So, keep these in mind when picking or making toys for them.

A small box full of shredded paper makes a great dig box that they can enjoy. (Get ready to clean up a mess!) Toilet paper rolls stuffed with hay make great chew toys. Some rabbits enjoy toddler toys like plastic keys, stacking cups, or things with bells on them. Others just love destroying a good game of Jenga, or ripping apart an old phone book. They tend to make a game out of anything you are doing with them. Honestly, they just love spending time

with you. They love when you interact with them, and my rabbits love for you to talk to them, brag on them. Honestly, they seem to know and understand when you say "Good boy" or "Smart girl." You'll sometimes get you an extra-high binky (see page 97) when they realize they did something you liked.

There are lots of toys that you can purchase online or in the pet stores, but honestly, making toys is a great way to limit costs and can be fun. This little activity table is made from untreated pine and sisal rope and can be made to any size appropriate for your rabbit. It is a great way to deter them from chewing on your nice furniture. Positive reinforcement will help them realize that this is their table and it's okay to chew.

If you see them headed for a table leg they're not supposed to chew on, a firm "No, no," is usually enough. If it isn't, gently tap them on the head and say it again, "No, no." They will usually run off. When you see them on "their table," respond with "Good boy!" or "Good girl!" Also, keeping a small jar of treats handy is a great way to reinforce this behavior. They will eventually chew away the sisal on their table, but it's easy enough to replace.

Hanging toys are also a fun activity for rabbits. If they have their own area, you can use a piece of twine and some clothespins to create a garland for them. Placing it just slightly higher than they are tall will cause them to have to raise up to get to the fun treats. You can hang herbs, kale, bells, sisal, appropriate twigs, or paper—so many options, and it is a great way to nurture their curiosity.

Another type of hanging toy that you will find in the pet stores contains pine cones. If you sterilize pine cones you find outside, they make great toys.

Sterilizing Pine Cones for Chew Toys

Rabbits love to chew on pine cones, and in the wild, they have plenty of them available. Undomesticated rabbits have already been exposed to many of the things that can be crawling on these pine cones and have likely developed immunity, but your pet rabbits are more sensitive. Once cleaned and baked, pine cones can be an excellent way to help maintain your rabbits' teeth naturally. Preparing them is simple, but it does take a few days to make them completely safe.

We cannot stress enough that you *cannot* safely give your rabbits pine cones directly from the yard without cleaning and treating them first.

Step 1:
Locate pine cones that are safe. This means that they are away from roads and edges of yards. Close to roads and parking lots, pine cones can have dangerous liquids, such as fuel, antifreeze, weed killer, and so much more that could prove very fatal. If you are gathering from the edge of a yard, the same can be the case. Fuel from lawn mowers and weed eaters, weed killers, fertilizers, etcetera, need to be considered. We prefer to go a ways into the woods, away from drainage routes, and collect pine cones there.

Step 2:
Fill a sink with warm water, and mix in 1 cup of white vinegar. Scrub the pine cones to remove any loose dirt, sap, and bugs that may be crawling on them. Allow them to soak in this water for at least 30 minutes. Make sure to rotate the pine cones around until all areas are underwater.

Step 3:
Rinse the pine cones really well to remove anything left on them, including the vinegar. Shake out the water and lay on a towel to dry. (You can place them in a dehydrator if you have one available.)

Step 4:
Once they have completely dried out, place them on a cookie sheet and put them in a 200- to 250-degree oven for approximately two hours, until they have time to completely open up. Stay close to check on them and turn them a couple times. You don't want to leave them cooking because they can burn easily if the oven gets too hot.

Step 5:
Allow the pine cones to completely cool down before you offer one to your rabbits.

Chew Mats

You can easily create a small mat or rug out of a bale of sisal, a large needle, and some thick thread. Start by coiling the mat and slowly bending it into a circular shape. After each circle, take the needle and thread, and secure it to the prior section. Keep circling until you reach the desired size. Make sure you never use any glue, as it could be toxic. Your rabbit will love sitting on this mat while they begin to chew it apart.

Toilet Paper Roll Toss

Taking an empty toilet paper roll or paper towel roll and stuffing it with timothy hay, dehydrated carrot chips, a couple raisins, some shredded paper, dried herbs, sisal strips, etcetera, makes a great distraction for rabbits who will have a long day at home alone.

Chew Bites

Random chew toys are easy to put together to entertain your rabbits. Simple baby blocks can be used with the sisal, pine cones, cardboard pieces, and so much more. Once all the chewable parts are gone, remake them again!

Bunny Bits

Make some new fun toys to leave for your rabbit sitter if you are going to be out of town for a few days. Each day, the sitter can give the rabbits a new toy to play with!

Rabbit Habits and Health

When you first get your rabbit, you are going to notice some slightly odd behaviors. Many of these are perfectly natural and are useful for rabbits living in the wild. It is amazing to watch your indoor, domesticated rabbits showing these instinctual behaviors. It shows how much is passed down through nature, rather than learned from other rabbits.

Digging and Smoothing

Whether it is the bedding in their cages, their blankets, your blankets, your jacket, a potted plant, or the edge of rugs, you will see your bunny determined to dig. They love to dig! They can dig for hours if they are allowed to. This can be aggravating or adorable depending on the moment.

Digging is a necessity in the wild. As social animals, rabbits will work together to create extensive underground tunnels that interconnect with each other. These communal living areas are known as "warrens." They also create many entrances and exits so that they have an escape route if needed. There even appear to be "living chambers" within the areas that they claim for themselves and their young.

As they dig these tunnels, they push the dirt out of their way, and eventually out of the tunnel. The process we see with our blankets appears to be a mimic of this behavior.

Rabbit Activity

Choosing to cater to instinctual behaviors will help your rabbit be happy! Here are a few suggestions to help the digging/smoothing urge.

Reinforcing good behavior is the only way to teach a rabbit. You cannot discipline rabbits in any kind of physical way—they will not understand. If you find your bunny is digging in the carpet of a certain area, or just has a drive to dig, create a "dig box" for him to enjoy! When you see him go to his "dig box," give a small treat! He will soon learn that this is a rewarded behavior.

A "dig box" can be as simple as a cardboard box with shredded newspaper in it for an indoor-only rabbit. If you sometimes take your rabbit into a safe area outside, create a small bed for him with untreated lumber and some soft, fresh dirt in it. Make sure that the dirt does not have any added fertilizer or vermiculite added to it, as these could make your bunny sick if it is accidentally ingested. You can usually find an organic topsoil that is safe.

Curiosity and Periscoping

Rabbits are very curious by nature. It is very easy to recognize a curious little bunny. You will see them stretch their neck and tilt and turn their ears in different directions to pick up a sound they hear. Their nose wiggles most of the time, and it is usually a signal for how interested they are in what they are observing. They remind me of a cat, in that they seem to navigate their surroundings several times a day to make sure everything is in its proper place, or at least the same spot it was in the last time they checked. They will be curious about what you are up to also. If you are in your favorite reading chair, having a snack, they will most likely come to see what you are doing. Keep some healthy nonperishable treats handy to give them.

Periscoping is a basic behavior you will probably see often, particularly when they are curious. Basically, they will stand and balance on their back legs. Normally, they are hearing something, or trying to see something, and they need to be taller. They will also stand like this if you are offering them treats. This is another example of a behavior that has developed and adapted from the wild instincts. You can imagine a wild rabbit hearing something way off, and raising up to try to see or hear it better. You can also envision a rabbit being down in their hole

when they raise up on their back legs so that only their ears and head are visible to potential predators. This is why the term "periscope" was used to describe the behavior. We love it when Sugar periscopes! It is just so cute! Often, he periscopes the room to see who is sitting in what chair! Periscoping is a great opportunity for a photo!

The Binky

A binky is a rabbit's ultimate way of showing how happy it is. The best way to describe a binky is a quick, bouncy little jump that a rabbit does when they are excited. They usually also twist their bodies while they are in the air, and do a quick little head tilt at the same time. This acrobatic display is hard to explain, but impossible to miss when you see it. Basically, the first time you see your bunny do this motion, you can be sure you are doing something right, and it just made them jump for joy!

Chinning

Chinning is a natural and territorial behavior. Directly underneath a rabbit's chin, there are scent glands. These scent glands are activated when they press them against an object, leaving their scent on items such as furniture, clothing, blankets, stuffed animals, you, basically anything they can reach. In rabbit language they're saying, "I chins it, it is mine now." Luckily, this scent is not detectable to humans, but other animals and definitely other rabbits will immediately recognize the smell and

the fact that the territory has already been marked. You may see your bunny chinning an item and think he or she is nibbling, so don't be too quick to assume so.

Licking/Nipping

Licking is a sign of affection from your rabbit. This is a grooming behavior that you would normally see with a bonded pair of rabbits. Grooming each other is a major way they show each other affection. The nipping is also a carryover from this. A rabbit may "nip" another one, working at something in their fur. In the absence of fur, it feels like a small bite to us, but you will learn the difference. They don't understand that it hurts you, so make sure that you don't ever attempt to discipline them. Remember, they think they are showing you love.

Humming/Honking

One of the only sounds your rabbit will make is a slight humming or honking noise. It is usually very low pitched, and you could easily miss it. Usually you will notice this behavior when they are licking you, mock mating, or they are around another rabbit that is causing them to be excited. They will also make this noise sometimes when they are circling your legs. All these signs are affectionate, and they do not make this noise in any form of aggression.

Bunny Bits

Rabbits actually make many different sounds that we cannot hear.

The Flop

The flop is one of the cutest and scariest things your bunny will love to do to you. The flop is actually a good sign. It means your rabbit is content, feels safe, and is relaxing. However, the first few times you see this, you will think he has passed out, or worse. Some rabbits react quickly to you calling their names, others will be so *blissed out* in their flop that they ignore you.

Thumping

The loudest sound you will regularly hear your rabbit make will be the thump. The thump happens when they are sitting, and they raise up just enough to slam their back legs against the floor. The force that they are able to use creates a surprisingly loud sound. Normally, your rabbit will do this when they are either scared, anxious, or even annoyed. In the wild, the thump was used to warn other rabbits that danger was approaching. It really is amazing how loud it is when Sugar thumps. He doesn't even look like he's moving!

Spraying

Spraying is one of the few very annoying things that rabbits can do. The easiest remedy for this is to have them spay/neutered in a timely fashion, before they ever begin to spray. Many people have a hard time with this, but I have been lucky. None of my rabbits were ever sprayers. We have seen them accidentally spray if something scared them or startled them, but it was never bad, and never intentional. For most rabbits it is a territorial claim on an area. However, you have to be careful and pay close attention to this. If they are all of a sudden urinating outside their area, missing their litter box or something like that, it can signal medical issues like urinary tract infections or bladder sludge. Spraying is an obvious spray that lands vertically on a wall, side of the cage, a piece of furniture, or perhaps even your leg. A puddle outside their area a couple of times is cause for concern and a visit to the vet.

Screaming

Honestly, we hope you never hear this sound. A rabbit only screams for a couple of reasons: extreme fear or extreme pain. If you do hear it, something bad is happening and you are needed immediately. A rabbit in fear for its life will sometimes scream. Something may be after her, chasing her, or something else may have happened that really scared her. The rabbit may also have a really bad and painful health issue going on like a painful case of bloat. Anytime you hear the scream, we will suggest you head for the vet. If it is health related, you need the vet's help immediately, if you have any chance of saving your bunny. Even if it is a fear-related scream, something has scared them enough that they could have a heart attack or go into shock.

Grab a blanket or a towel and head for the vet. If possible, have someone drive you so you can hold your rabbit against you, wrap them in the blanket to keep them warm, and give light tummy massages. Call your vet's office to let them know you are coming in with an emergency. Hopefully, they will make room for you so that you can come straight in, and they may have pain meds ready to go. This is another reason why it is so important to have a great relationship with your veterinarian.

Building a Rabbit First Aid Kit

First off, we will say that rabbits get sick fast, and have to be treated fast. By the time we notice the symptoms, a rabbit may have already been dealing with an issue for a while.

No matter what the issue—a wound, respiratory illness, or digestive crisis—you need to go to a rabbit-savvy vet ASAP; these suggestions of what to have on hand for emergencies are only to supplement vet care.

It takes an initial investment to get this ready to go, but it is so worth it when you realize your bun is sick. If you can't get it all at once, I would try to start buying a couple of the items every few weeks and adding to your kit until you have it complete. We definitely didn't have all this starting out; it has taken years of trial and error to get our kit like it is today, and some of these things may not work for you. This is simply a suggestion—you may need to add many items to this—but this is a good place to start.

We are going to go through a few of the major suggestions to help you understand why we included them, but most are self-explanatory.

Rabbit First Aid Kit

1. The first thing we would suggest having in your first aid kit is **Vet Rx Oil for Rabbits**. This product is wonderful for rabbits with colds/allergies, and it can also be used to treat ear mites. Plus, we love the fact that this product is all natural.

2. **Little Remedies Gas Drops (Simethicone)** are used to treat bloating and other digestive issues.

3. A **stethoscope** is a great addition because you can listen for bowel sounds, in addition to having it to help detect respiratory issues.

4. When they get sick, keeping their tract nourished and hydrated is important. We have **Oxbow Critical Care** in the freezer just in case. This is a finely ground feed that is nutritionally complete and is easy on their system. It can be easily mixed with water and fed via syringe. All rabbit owners know the fear of stasis. This happens when something like wool interferes with the intestinal tract of your rabbit, and causes them to stop eating/drinking/pooping. When this happens, you need to get to the vet, ASAP. The products we are suggesting here are only to help prevent and restore balance to their intestines. Please always go to the vet immediately.

5. The final product that we are suggesting for digestive issues is a **heating pad** that was created for animals. Snuggle Safe provides heat safely, which is recommended to help them battle stasis.

6. Whether you are treating a wound like sore hock, a broken toenail, or a bite, once again, start at the vet. However, we keep **Vetericyn Wound Care** in our kit. This spray helps clean the wound and can help prevent infection.

7. We keep **styptic powder** on hand to stop bleeding. Whether it is a simple scratch or a toenail we trimmed too much, styptic powder can stop the bleeding immediately. (In a pinch, you can always use cornstarch.)

8. **Dried chamomile** is an herb that has amazing healing abilities. Not only will it help soothe the nerves of a rabbit with anxiety, but it also has natural antibiotic properties. So, we offer this daily to our rabbits, free

choice. But we also keep it to treat any type of eye issues, such as runny eyes or infection.

9. **Vet Wrap** is a product that we try to keep on hand for emergencies. It is great to wrap around a wound because it sticks to itself, but not to skin or fur. You just need to make sure they cannot chew on it.

10. **Vetericyn Eyewash** and **Vetericyn Ophthalmic Gel** are essential for cleaning and treating irritated eyes.

11. A pediatric **rectal thermometer** will also be a great item to have in your kit. During one of your healthy rabbit appointments at the vet, ask them to show you how to correctly and safely take their temperature.

Preparing Your Rabbit for Surgery

If your rabbit has to have surgery, it is a very concerning thing for both you and your bunny, but there are a few things you can do that will help, and a few things you need to know. The most common surgery, of course, will be the spay/neuter, but all surgery preparation is basically the same. A good rabbit vet is the most essential thing you need. Any good rabbit vet will prepare you for everything prior to surgery day.

It is essential that your rabbit not be fasting prior to surgery. This is only necessary in an animal that can vomit, and a rabbit is unable to vomit. Fasting is very dangerous for them as it can cause many digestive issues, especially when combined with the stress of surgery. So, if any vet advises you to have them fast prior to surgery, you need to find another vet.

Before heading to the vet's office, prepare your rabbit's food. Take a bag of her pellets, hay, and a selection of her favorite fresh foods. In addition to this, take her water bottle

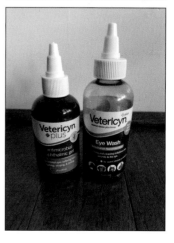

or bowl and her blanket. Basically, you want her to have everything she needs to feel comfortable, and to be prepared to give them food and drink immediately after surgery.

If your rabbit has a bonded friend that they are always with, most vets will suggest that you bring both rabbits in so the stress of the stay will not be made worse by being away from their bond-mate. This may also help reduce the chance of them rejecting each other when you return home.

If it is at all possible, try to schedule surgeries on a day that you can stay with your rabbit at the office. It will be reassuring for them, and it will help them recover faster. You should be able to stay with them until they are asleep, and be in the room with them as they wake up. They will look to you for comfort since they will be in pain that they do not understand. If they wake up from surgery alone, in a strange place, and in pain, they may panic and cause other injuries.

Before you head home with your rabbit, make sure you have all the medications you may need, and all of the items you brought to the vet with you, i.e., water bowls/bottles, blankets, etcetera. Make sure that you get pain medications for your rabbit. Being in pain will cause a great deal of added stress, and it will cause them to not eat much or at all. Once again, not eating will cause even more issues, so I would supplement with Oxbow Critical Care for several days after bringing them home. Keep in mind that you may not be able to tell that they are in pain; since they are a prey animal, they have learned to hide this.

How to Tell If Your Rabbit Is in Pain

A rabbit who is in pain will work very hard to hide the fact that they are hurting. There are several different ways to determine if they are in fact in pain. Of course, if your rabbit has had surgery, assume they are hurting for the first couple of days and give the prescribed pain medication to them as directed by the doctor.

The best way to determine if your rabbit is in pain will be to watch them and see if they are acting differently than normal. Are they sitting in a hunch? Are they hiding in a corner when they are usually out running around? Are they sleeping a lot more than usual? Do you hear a teeth grinding sound? Are they breathing rapidly? Are they suddenly aggressive? Are they eating? Will they come to you for a treat? Have they used their litter box today? If you notice anything unusual, you will need to begin monitoring your rabbit closely, perhaps call the vet, and be prepared to head for vet's office if your rabbit doesn't begin to improve soon.

THE BUNNY BURRITO

We have mixed emotions about the bunny burrito. Sometimes you may need to give your rabbit medicine, and generally, they don't like to be forced to do anything like this. So, to help, people will carefully wrap their rabbit in a towel or blanket, leaving only their head exposed. This helps to be able to give medications. This may be very necessary, but you must always be careful when trying to force your rabbit to take meds or drink something. This should only be done if they are sick, and it is absolutely necessary.

I had to give Sugar medicine within a few months of getting him. He didn't trust me yet, and he didn't like me trying to force him to take something. The bunny burrito approach made it difficult and frustrated us both. The next time I had to do meds for him, it wasn't any different. He hated it. After much trial and error, I decided to try something. I simply got his meds in a dropper, and I put him in his carrier. I put the dropper near his mouth, and in just a couple tries, he willingly took the dropper. This is so much safer, with no chance of aspiration and no stress. What he hated was feeling confined and forced. He will take his medicine, Critical Care, and water using this approach. So, I suggest trying this first, and only doing the bunny burrito if you absolutely have to.

The Most Common Rabbit Illnesses

This is just a basic listing of things you may encounter in the first year of having a rabbit. More than likely, you will not have these issues, but all of these will require specialized veterinary care quickly. There are also many other things that can happen; please do your research and have a rabbit-savvy vet ready in advance. You don't want to have a sick rabbit and be attempting to find a vet on the same day.

Bunny Bits

A healthy rabbit's normal body temperature can range anywhere from 101 to 104 degrees. Anything higher or lower is signaling an issue that needs to be treated.

HERPES

Herpes is considered to be a zoonotic illness, meaning that it can pass between human and animal. If a rabbit contracts herpes, it is fatal. A huge proportion of the human population carries this virus, which often manifests as cold sores and fever blisters. If you have an active herpes infection of any kind it is vital that you keep your hands extremely clean while you are handling your rabbit, their food, their bedding, or anything else. You must also keep your face away from theirs. They can be infected through their eyes, nose, or mouth, if you touch your sore accidentally, then rub them, kiss them, or allow them to lick your face. You must be vigilant as long as you have any visible sore. They can be treated for the virus just as you can, but never cured, and it is often misdiagnosed. So, if you have possibly contaminated your rabbit, make sure that you let your vet know so that any future illnesses can be considered.

SNUFFLES

The term snuffles is used to describe a wide variety of cold-like symptoms that your rabbit can have. They can have chronic runny eyes, runny nose, sneezing, etcetera. This is normally caused by a bacterial infection, *Pasteurella*, starting in the tear ducts. This is a very serious health condition that will need to be immediately treated with antibiotics to stop the spread.

Rabbits who have dental issues are prone to getting snuffles, but a great way to prevent this is to keep their environment clean and free of fumes, and feed them a high-quality diet that is high in fibers from the hays they are eating.

FLEAS/MITES

Flea infestations are very dangerous for rabbits, and your indoor rabbit can get fleas just like an outdoor rabbit. It is far easier to prevent your rabbit from getting fleas than to attempt to rid them of them afterwards. Most flea treatments are dangerous for your rabbit, so again, prevention is best.

Tricks for keeping fleas away:

- Keep their environment clean.
- Wash bedding regularly—at least once a week.
- If your rabbit is outdoors, change their hay/shavings weekly.
- Use flea prevention on any other animals, such as dogs and cats, that they are around.
- Whether indoor or outdoor, diatomaceous earth is a safe way to treat the area.

Fleas can be picked up from other pets or carpets, or your indoor rabbit can contract them from even a quick trip outside.

Symptoms of Fleas:

- Excessive licking, scratching, and chewing
- Scaling of the skin
- Hair loss
- Bites on skin
- Flea dirt in fur
- Anemia
- Bacterial infections

Let's start with what not to do. Do *not* use Frontline on your rabbit. Many over-the-counter flea powders, flea shampoos, and flea collars are also not safe. Remember that rabbits are self-groomers. Anything that you put on them they will ingest as they are grooming. Shampoos are never an option, because bathing is dangerous for them anyway. This causes poisoning and can be fatal. Look at all ingredients; if anything you are considering using contains fipronil or permethrin, then it is not safe at all.

If you discover fleas, we recommend going to the vet and getting a prescription for Advantage. This is the only flea medication that we are going to recommend because it has been proven to be safe more often than not in rabbits. Also, a vet will weigh your rabbit, and give you the exact amount that you need to give, and tell you how often to give it to totally get rid of the fleas. The vet will also show you the proper procedure for application. More than likely, it will take more than one dose to rid them since fleas can be killed, but the eggs they have laid will hatch later. You can also ask your vet to give you refills once you know the correct dosage.

If you catch it really early, and your rabbit is calm, then you can attempt to treat them more naturally. Purchase a flea comb and begin grooming them daily. If you have time, twice a day will be better. A small amount of diatomaceous earth (de) can also be used to help get rid of the fleas. Simply use the flea comb and dip it in the de powder, then gently comb through. As you are grooming, you can actually watch for fleas. If your rabbit is calm enough, you may be able to get the fleas off. However, don't wait too long. If you are still seeing fleas or the "flea dirt" in their fur after a couple days, then go ahead and get the meds from the vet and stop it before it gets too advanced.

Fleas carry all sorts of bacteria, and this can be transmitted to your rabbit through a bite. Also, mites and ticks can carry many of the same symptoms as fleas, but they are treated in different ways. So, once again, it is always safer to head to the vet. Once you have learned about these issues from your vet, they may be willing to give you refills without a visit. You just need a good relationship with them!

Mites

There are three major types of mites that affect rabbits: mange mites, fur mites, and ear mites. Mange mites are tiny and cannot be seen. Fur mites can be seen and are two-tone in color. Mites can be transmitted from another infected rabbit, infected blankets or bedding, hay, or even a trip outside. A lot of the symptoms of mites are similar to those of fleas. The major difference is that mites burrow into the skin of the rabbit. They can create sores, which can become infected, and many times you will see what appears to be dandruff accumulating on them. This is a definite sign of mites.

Ear mites are tiny and they accumulate in the ear canal. This causes extreme irritation and infection. The sores in the ear, cankers, can become hugely swollen, inflamed, and infected. Many neglected rabbits come into the shelters in this condition.

All mite cases need to be treated seriously and quickly. A vet can prescribe a correct dosage of something to eradicate the mites. This may need to be given multiple times over the course of several weeks. You should also discuss the need for an antibiotic if there are sores present.

Myxomatosis

Myxomatosis is a viral infection that is a huge issue in Europe and becoming more common in North America. This infection is transmitted to rabbits only through mosquito bites, flea bites, or fur mite bites. There is no vaccine available in the United States, and unfortunately, it is usually fatal. Protect your babies from these tiny pests, and get them treatment as soon as you see any of this.

Bloat—"Gas"

Bloating is a very painful condition that many rabbits will experience at some point in their lives. Knowing your rabbit's daily routine and personality is important when identifying health issues early. A rabbit who has suddenly become lethargic, is sitting hunched over, and shows no interest in activity or food has something going on. Start by feeling their tummy. Is it hard and rounded? Does your rabbit flinch or move away as you are trying to feel it? If so, they may be showing the early stages of GI stasis or bloat. At this point, call your vet, and see what they suggest doing. If you can't get in touch with them, it is late at night, or something like that, below are some suggestions to help you get through the first few critical hours. Always remember, if you have been gone all day, you don't know how many hours this has already been going on.

Many things can bring on bloat. A new food can easily trigger an episode. Perhaps they have chewed something they shouldn't have, such as carpet or a piece of furniture. A particularly stressful day can even upset them enough to trigger it.

The problem with bloat is that the attack comes on without much warning, is horribly painful, and can kill them in just a few hours if not treated. Another issue with bloat is that it can be caused by a blockage. Once again, if you administer the treatment, and see no improvement, you should head for the vet as soon as possible.

This is one major reason that the first aid kit is important. Having what you will need on hand will save you much anxiety and a panicked run to the store. One of the items in the kit is simethicone, or infant gas drops, which can be purchased at any drugstore. Try to find the dye-free option that has no flavoring added. These drops can break down the gas bubbles that are developing in your rabbit's stomach/intestines.

When a rabbit begins to get sick with this, their body temperature will begin to drop. Immediately call your vet; they may want you to come in immediately, or they may call emergency meds into your drugstore. While you are getting medicine together, or having to run to the store, you will need to keep your bunny warm. There are wonderful microwavable pet heat pads that are perfect for issues like this. If you have purchased one in advance, heat that up and put your rabbit on it. If not, you can create a homemade one by using a drink bottle, like a plastic soda bottle, and putting really warm water in it. Then, wrap it in a towel, and set it next to your bunny. It will not hold the temperature long, but it will work temporarily.

Holding your rabbit during this is also a good option. Tummy massages are also an excellent way to help your rabbit through this. As you are holding them, you are warming them with your own body heat, and you can easily massage their belly. You can gauge the pressure by their response. If it hurts, they will flinch and move away. Keep this going, and continue to offer water and hay options.

You will need to give your rabbit a dose of simethicone, approximately 1 cc, at the first signs of bloat, and repeat

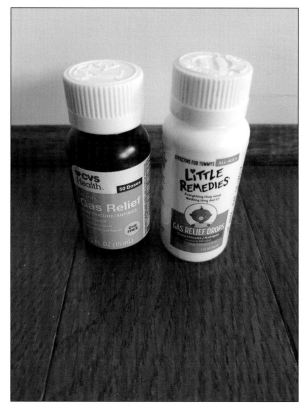

approximately every hour for the first three hours or until your rabbit returns to normal activity. Do this carefully and slowly, so they don't aspirate the medicine. Sometimes you can see improvement in just one dose, but generally, I repeat the dose for a couple hours anyway. Between the added heat, the simethicone, and the belly massages, you should begin to see improvement. If you have a feeding syringe, you can offer water directly if they aren't showing any interest. However, you must be careful not to force it, as they can easily aspirate and that will make the situation much worse.

If you aren't seeing much improvement within the first few hours, call the vet again. If no improvement at all is noticed, then head for the vet's office, taking the heat pad with you. The vet can give fluids, IV meds, pain meds, and more. It is always better to be safe than sorry, and the reassurance of a vet will help you realize you are doing the correct steps. Just don't wait too long; any sickness in a rabbit is an emergency because they get sick and become critical very fast.

GI Stasis

GI stasis presents similarly to bloat because once stasis has set in, they will develop bloat also. Gastrointestinal stasis is a very dangerous, often deadly condition that develops when the intestinal tract begins to slow down and eventually stops moving completely. It is extremely painful, and develops very quickly. Immediately head to a veterinarian. Your rabbit will need medications to overcome this issue. Depending on how advanced it is, they may even need IV meds to jump-start their intestinal tract again. If this happens, once you have dealt with this issue, you will know how to handle it better and will better recognize the signs your rabbit will exhibit when they are getting sick. Also, ask your vet to give you emergency meds to keep at home in case it ever starts again. This way you can immediately start them on the medicine.

I am a student of herbs. I have spent many years growing and using herbs. I view herbs as a preventative, an immune system builder, and an overall health booster. However, things will happen, and in an emergency, I will take my animals to the vet. I will give antibiotics if recommended, and I will readily admit that a chemical medication saved my rabbit's life many times.

I also firmly believe that herbs are underused in our culture, and I believe that medications and antibiotics are overused. However, if herbs were all that were needed, then chemical antibiotics and other medications would have never been studied or created.

A few weeks ago, I noticed one evening that Sugar was just lying around. Now, this isn't necessarily out of the ordinary. Every rabbit is allowed a bad "hare" day once in a while. So, I gave him some fresh hay, romaine lettuce, and fresh pineapple, and I went to bed. The next morning, I came down, made my coffee, and started for the den where he sleeps. I knew something was wrong before I even got in there. So here was what I found:

1. Sugar always starts thumping when he hears me in the kitchen making my coffee. He cannot wait to get out in the mornings, but this morning, he did not.
2. When I got to his room, he was still lying down.
3. When I opened the gate, he didn't get out.
4. He had not eaten any of his lettuce, hay, or pineapple from the night before.
5. I remembered he was rather lazy the night before.

6. I got him out, and he just stayed on the couch with me. He had no desire to play.
7. I saw no evidence that he had used his litter box.

At this point, we knew something had gone horribly wrong with Sugar's system. We left immediately for the vet's office. We couldn't get an appointment till 3 p.m., so we just went, signed in, and waited in the car. We told them it was an emergency, and we would wait till she could work me in.

About an hour later, she worked us in and diagnosed the dreaded gastrointestinal stasis. She did say that it appeared to be in the very early stages. She said that I had done nothing wrong, his weight was perfect, his diet was perfect, and he was well groomed. She said it was just one of those things that happens. What is so sad, however, is that she said most of the time people do not get their rabbits in soon enough to save them. She said another 24 hours and we would have had a different outcome. Rabbits cannot survive for very long without eating, and with no fecal output, his system had blocked up. Gastrointestinal stasis causes their intestines to stop up. The first sign is that their poops will change in size, and sometimes be connected to each other by strings of hair. At this point, you can cut up fresh pineapple and papaya to counter this. If that doesn't work, fecal output will continue to decrease in size and slow down to an eventual stop. At the same time, the rabbits will stop eating, and start laying around and acting lethargic. The night before Sugar got sick, he was lethargic, but did want his favorite treats. By the next morning, he wasn't eating at all.

Within a 24-hour period, he had gone from a bouncy, healthy bunny to an extremely sick one.

We left the vet with a week's worth of a liquid medication called Reglan. We ordered Oxbow Critical Care and had it the next day. She showed us how to safely give him his medicine and the Critical Care by syringe, and told me to go to the store and get baby food—mashed pumpkin and apple. We were to make him eat through the syringe about every hour or two during the day. In addition, she said to get pineapple juice, and make him drink that to help dissolve any hair he may have in his gut.

The Reglan made him extremely drowsy, but he started using the litter box the day we started the medicine, and his droppings began at about ¼ their normal size. After just six hours with the medication, he began eating a little hay on his own. It took about a week before he was his normal, happy self again. During this week, we regularly were syringe feeding him extra Critical Care and water to make sure he was rehydrating properly. I was with him constantly—I even took him to work with me.

We learned a valuable lesson though. We would rather pay an unneeded vet bill than ever risk Sugar's life. Twenty-four more hours, and he would have either died, or needed surgery that he may not have survived.

My advice is this:

1. Trust your instincts.
2. Know your animals.
3. Provide fresh clean water, hay, and pellets daily, as well as fresh lettuces and some fruit like pineapple. Hay should make up a major part of their diet; however, we provide unlimited hay and pellets. Plus, they get the fresh, organic vegetables, lettuces, and some fruit twice a day.
4. Keep their living conditions clean. If we didn't change litter boxes every day or two, it would be impossible to tell whether he was using his box.
5. We believe the herbs keep my chickens and rabbits healthy and happy; however, we will not rely solely on them. Once sickness sets in, we will go to a vet.
6. Finally, keep emergency medications on hand. Your vet will normally be willing to give you these in case of an emergency. A well-stocked first aid kit will be a true lifesaver.

We believe fresh/dried herbs daily as a preventative and immune booster are a powerful tool for everyone, including our pets.

However, in an emergency, we will, and we will always advise you, to take your animals to a vet to get diagnosed.

Bunny Bits

An early warning sign that a rabbit's digestive tract is getting clogged are poop pearls.

Unfortunately, once your rabbit has had GI stasis, they seem to be prone to it. You will learn to watch for it, and if you begin medications at the first signs, you probably won't have such a hard time treating it. Sugar has had it several times, but never to the extent that he did that first time. It is always scary; just stay calm, treat them quickly, and be ready to head to a veterinarian if they don't immediately respond, even if you know how to normally deal with it.

Head Tilt

If you walk into a room and see your rabbit sitting on the floor with their head tilted to the side, and they are unable to straighten it, you are dealing with a case of head tilt. Head tilt is a very scary illness that comes on your rabbit quickly. There are many things that can cause head tilt, but the most common is an inner ear infection. If this isn't treated immediately, the head tilt can become permanent, causing your poor rabbit to have to relearn how to walk, eat, and everything else. A simple round of antibiotics given immediately can reverse the head tilt. The veterinarian will need to examine them, and even do a blood test to rule out other more serious conditions.

There are many symptoms associated with head tilt. The most common, of course, are dizziness and the resulting rolling and spinning from that. Some rabbits can have seizures and even lose the ability to walk.

If your rabbit is on the medications, you must give them at the correct time and dosages. With proper care, you may start seeing the head tilt improving. In the meantime, you will need to protect them. If they are rolling or spinning, you should roll a couple towels to help prop them up. You will also need to ensure the fact that they are getting enough to eat and drink so that this doesn't trigger a gastrointestinal issue also.

E. Cuniculi

Another cause of head tilt is *E. cuniculi*. This is a parasitic infection that attacks a rabbit's spinal cord and brain. Most rabbits are carriers for *E. cuniculi*, but it only shows up in a small fraction of rabbits. *Encephalitozoon cuniculi* is the parasite that causes many issues in rabbits. Many things from stress to another illness can actually activate the parasite, and it can present in many different ways. If they weren't born with this parasite, they can contract it by breathing in the spores, coming into contact with another rabbit who is carrying it, or surfaces/foods that have been contaminated. If it attacks the eyes, they can develop cataracts quickly. There can be paralysis, kidney issues, head tilt, seizures, and other issues. The good news is that there is a treatment for this, and if it is started at the onset of symptoms the chance of recovery is very good. A blood test will be done to confirm the diagnosis, and medication should be started immediately. Panacur, which is an anti-parasitic, is usually the first line of treatment, along with an anti-inflammatory. Another important part of the treatment will be to disinfect their living areas. This parasite is able to live for nearly a month, and all surfaces should be disinfected to remove the chance of reinfection.

Sore Hock

Sore hock is an extremely painful condition that rabbits can get for many reasons. The most common reason is being kept in a cage with a wire floor. If rabbits have no way to get off the wire, then their fur will eventually wear down. Sore hocks are actually pressure-related issues caused by their body weight where their feet can become raw, and if not treated, the sores can become infected and dangerous.

The trick to treating this issue is to catch it very early. As soon as you see the fur thinning on their feet, you can begin to treat it. If there isn't a sore, you can attempt to treat it yourself, but if the skin has already broken, it would be better to get some medication to help keep it from getting infected. Your vet will prescribe a safe medicated ointment to ensure that infection doesn't set in.

If the skin hasn't broken or become a sore you can attempt to protect their feet and allow the fur to naturally regrow. The most important thing is to figure out why it happened. If they are on wire, you need to fix this issue. Are they on a plastic bottom cage all day or on wood floors? The inability to get a solid grip on the floor may be wearing their fur off. Make sure they have a soft surface to stand/lay on. I use baby blankets in the bottom of their cages and nap areas. Make sure they aren't in wet conditions either. If their bedding, whether it is shavings or blankets, gets wet, change it immediately. Are their nails too long, causing them to put weight on their heels? Are they becoming overweight? Basically, discovering the issue is the first step in solving this problem. Once the cause is fixed, it will be easier to promote fur regrowth.

You can use infant socks in a pinch to protect your rabbit's feet while treating for sore hock. Carefully pull the sock over its foot and mark the point where the foot bends at the top. Take the sock back off and use a pair of scissors to cut a small hole at the mark to make it easier for their foot to naturally bend.

Bunny Bits

We keep a roll of Vet Wrap in our rabbit first aid kit. This wrap is great because it sticks to itself, but not to skin or fur. You can use a tiny bit of vet wrap to gently hold the sock at the top so your rabbit can't easily kick it off. Just be sure not to attach it too tightly.

Warbles

Whether your rabbits live inside, in the garage, or outside, please keep their litter boxes clean. If you have an outdoor rabbit and don't use litter, please keep the area under their cage raked up. The importance of clean living conditions cannot be stressed enough. Sitting in urine can easily degrade a rabbit's fragile skin very quickly. Besides this, gnats and flies are immediately attracted to the dirty cage bottom, litter box, or even the ground under the cage.

The problem with this, besides the obvious, is the fact that rabbits are highly susceptible to a condition known as warbles. Warbles is a condition where a fly or gnat lays eggs in wounds, scratches, or any moist area, such as the eyes or even the anal area of any larger animal. Once this happens, the eggs hatch and they begin to actually eat the flesh of the rabbit. It is a horribly painful condition that usually ends in death, or at least an extremely high vet bill. This is 100 percent preventable with clean living conditions and the attention of the owner. Also, having a fan blowing on the cage, whether it is outside/ inside or in the garage, will keep the flies and gnats from being able to land on the rabbit. Indoor rabbits can also have issues with gnats, but only if their cages are dirty enough to attract gnats into the house, or if the house already has a gnat issue. These litter boxes make keeping their cages clean so easy, and rabbits are extremely easy to train.

Remember that animals who are kept in cages are totally dependent on us to provide everything they need. Be responsible and keep them healthy, happy, clean, and safe. They will reward you in many ways for providing for their needs.

Heat Exhaustion/Heatstroke

Summertime is dangerous for rabbits for several reasons, especially outdoor rabbits. Can you imagine being confined to a small cage, in the summer heat, with a fur coat on?

A rabbit does best when the temperature is 60–70 degrees. Heatstroke and exhaustion can begin being experienced in temperatures around the upper 80s, especially if the humidity is also high.

My first suggestion is to take any rabbit that appears to be suffering from heatstroke, heat exhaustion, or warbles to the vet, immediately.

Make sure your rabbit's outdoor cages are not in the sun! Make sure they have a full bottle of fresh water, fresh timothy hay, and some romaine lettuce and herbs, or some cool treat. Another really good trick is to put a fan on their cages, and put a bottle of frozen water in the path of the air. This helps keep the air moving around them, and will greatly help keeping them cool. My indoor rabbits love to sit in front of a fan. The fan also limits the ability of flies and gnats to land on your rabbit and their food.

All these may help keep your rabbit from suffering from heat exhaustion. If you notice your rabbit seeming lethargic, breathing fast, not eating, or any odd behavior, move them to a cooler location and provide fresh cool water. Add some mint if you have it available—mint helps bring down their body temperature.

In extreme cases, use a cool washcloth with a piece of ice in it to wipe their ears. If they appear to have had actual heatstroke (they can't stand up, are unconscious, have a fever, or are refusing to eat/drink, etc.), take them to a vet immediately. Rabbits with a health condition will deteriorate very quickly, and many times the only chance for survival is immediate vet care.

If you have outdoor rabbits, on extreme days, we would offer an additional water bottle or water dish that has electrolytes in it. Never fully replace an outdoor rabbit's water with this, however—rabbits are so picky that they may refuse to drink it, and end up severely dehydrated. (You can make your own electrolytes using water, sugar, and salt (see page 118), purchase an electrolyte pack, or even keep a bottle of Pedialyte on hand for emergencies.

Believe it or not, my indoor rabbits suffer from the heat too. We place a frozen ice pack in their cages every morning, and a new one every evening. They often lie on them, or lie on one of the air-conditioning vents.

We purchased several ice mats and ice brick packs from the dollar store, then made a cover to slip them down into, to ensure that the frozen pack wouldn't damage their skin, and that they couldn't chew into the ice mat.

Bunny Bits

Did you know that rabbits enjoy sitting on bricks or tile during the summer? Tiles are much cooler than wooden floors or carpet, and if you dampen the tiles or place a frozen bottle on them, it is even better.

Rabbit DIY

Electrolytes are a great thing to have on hand, and I always recommend keeping Oxbow Critical Care in your freezer for emergencies. Critical Care can be watered down, and it contains electrolytes and many beneficial vitamins for recovery. However, if you don't have any handy, you can always mix your own electrolytes with items that you probably have in your kitchen.

Homemade Electrolytes

1 cup warm water
2 teaspoons sugar
$^{1}/_{8}$ teaspoon sea salt

Mix these items really well, and place it in the fridge to cool down. Hopefully, you have a medicine syringe in your bunny first aid kit, which you can use to get them to start taking this a little at a time. You probably won't have to force them, because this will have a sweet taste that they will like. You can also offer this in a bowl and they may drink it on their own.

Shock: Scared to Death

Yes, a rabbit can be scared to death. What this actually is caused by is either a heart attack or shock. When a rabbit goes into shock, their circulatory system changes, and there is a drop in blood pressure and body temperature. Without proper treatment, they will not make it. It isn't a common occurrence, but it can happen, and we have heard many stories of people who believe that is exactly what happened to their rabbit. It seems to be especially prevalent in outdoor rabbits, but this is perhaps due to the amount of unusual things they can encounter. A perfectly healthy rabbit that is hopping around their enclosure, or perhaps your backyard, can be frightened by any number of things—a lawn mower, a leaf blower, a weed eater, a hawk swooping down on them, a dog appearing and chasing them, a car horn, a house alarm . . . so many possibilities. What actually appears to happen, is that they become so frightened that their fight-or-flight mode kicks in, basically causing a heart attack or causing shock. Because they are prey animals, everything in their world can be perceived as a life-threatening situation. Just be aware of any "new to them" activities you may be doing. If you are mowing for the first time in the spring, make sure someone is there to comfort your rabbit the first couple of times so that they learn that this loud sound doesn't have to be scary. I would even offer some calming dried chamomile treats for them to munch on prior to your noisy activity. Just try to imagine what life seems like to them. You will not be able to predict every possible scenario and protect them from it, but you can help them be calm and learn to become trusting, and this will go a long way toward helping them deal with anxiety.

In addition to fright causing these issues, rabbits can go into shock from an injury and blood loss. If you suspect any of these issues, you need to do two things. First, heat up their warm pad, or make a hot water bottle, and second, head for the vet. The loss of body temperature is one of the first signs, and you can begin to correct this on the way to the vet. The vet will administer IV fluids, and if it is indeed shock, then the combination may help save your rabbit.

Bunny Bits

The first few times you do a loud, new activity that your rabbit isn't used to, i.e., vacuuming, mowing, hammering, etcetera, have someone else do it first, while you sit with your rabbit. Hold them close, rub them, and speak calmly, telling them it is okay. This may help them learn that the sound shouldn't be associated with fear. It may take several times before they get totally accustomed to it.

Urinary Tract Infections and Bladder Sludge

Rabbits can get a urinary tract infection just like humans can, and it can be for many of the same reasons. If you notice your rabbit appears to have become incontinent, it is time to head to the vet. UTI isn't extremely common in young rabbits, but it can happen for several reasons. Here are the signs to watch for if you suspect a UTI:

- Frequent visits to the litter box
- A thick, brownish colored urine (can be light brown or dark)
- Bloody urine
- Incontinence

To help reduce the chance of UTI:

- Make sure your rabbit always has plenty of fresh water.
- Help your rabbit maintain a proper weight.
- Make sure your rabbit gets plenty of exercise.
- Limit confinement—they need plenty of space and free time.
- Don't allow too much calcium in their diet. Spinach and parsley are both high in calcium. Too much can cause stones, sludge, and other urinary issues. You can give it to them, but in limited amounts.
- Keep their environment clean.

Untreated UTI can get very dangerous, very fast. A veterinarian will need to prescribe the correct antibiotic and dosage for you to effectively treat this. In addition, they may need a mild painkiller.

Bladder sludge is a condition that rabbits can have if their calcium intake is too high. It causes a sandy-textured urine that is extremely painful. All the symptoms of the urinary tract infections can apply to this, but this can be a chronic condition. Once again, it generally doesn't affect younger rabbits, but if their diet has become too rich in calcium, it can happen. Just watch the foods you are offering and make sure not to give them several high-calcium items in the same day. Bladder sludge is another condition that you will need to visit the vet to learn how to treat. This is not something that can wait; treatment of anything that can potentially cause a blockage or an infection is a priority.

Bunny Bits

Freeze-dried cranberries make a great treat, and they may be beneficial for the urinary tract health of your rabbit. Try to use freeze-dried, organic cranberries so there are no added sugars.

Splay Leg

Splay leg is a condition that a rabbit is usually born with, but it can also be developmental, depending on the conditions they were born into. It is easy to detect because a leg or even all legs can be stuck out at an angle ranging from 45 to 90 degrees. It may not initially be visible, but can develop steadily if a rabbit is kept on a slippery surface. This cannot cause splay leg, but it can force joints that are weak to develop quickly into splay leg.

Splay leg is a disability, and a rabbit with this condition will need specialized care, probably for life. By the time you are able to get a rabbit, they should be at least eight weeks old, and more than likely, this condition would be very visible at this point. However, if you notice your rabbit standing strangely with one or more legs sticking out straight from their body, having trouble walking, one leg slipping away as they are standing, or anything like this, head to a vet. Your rabbit could simply have an injured leg, or it may be the beginnings of splay leg.

Bunny Bits

If you have a rabbit with severe splay leg or even a paralyzed bunny, there are companies online that make specialized carts so that your rabbit can move around on its own! Rabbits with these conditions can live happy, healthy lives with a patient, loving owner, and a lot of TLC.

Weepy Eye

Many things can cause a case of weepy eye. A new hay may be dusty, an eyelash may be poking them, something new in the house could be irritating, or it could be the beginnings of snuffles or *E. cuniculi*, teeth problems, or even an eye disease. Most likely, it will be something simple, but if it doesn't begin clearing up in a few days, you will need to have it checked out.

Bunny Bits

Try using a warm washcloth to clean up weepy eye for a few days, and see if this will help.

Vetericyn makes an eyewash and an ophthalmic gel that should be part of your first aid kit. Both of these items have antibacterial qualities, and if used quickly, they can help stop weepy eye fast. You can also make a compress out of a warm bag of chamomile tea. This will be very soothing to the eye.

If, after a couple days, you aren't noticing any improvement, head to the vet. More than likely, they will check their eye to make sure a tear duct isn't clogged and that there is nothing in their eye. If not, they will probably offer antibiotic eye drops first to see if that will clear it up. If it doesn't, they may want to do further testing.

Caring for Your Rabbit during Vacations and Holidays

Everyone enjoys getting away for a few days, and when it comes to having pets at home, you automatically know you have to consider their needs. Rabbits are just like anybody; some of them love to go somewhere new, and others prefer to stay home. The only way to know which your rabbit is will be to take them on a trip with you somewhere. Prior to leaving, make sure the place you are staying will welcome your rabbit. It has been great to see that many motels, restaurants, and shopping areas have become open to you bringing your pet in with you, rather than risking leaving them in the car. Just have a nice pet travel bag that you can easily carry in places with you, and line it with a pee pad, just in case.

Car Trips with Your Rabbit

Basically, traveling with your rabbit is like traveling with a baby. You have to bring everything you think they might need—a travel bag, their cage, blanket, food, water, first aid kit, and anything else they may need to keep them safe and happy. (Once you have done this a few times it gets easier.) We prefer to allow them to ride in their cage, because

Bunny Bits

Going on a few short car rides will be a great way to help your rabbit become accustomed to the sounds and movements inside a car, prior to a long trip.

there is a lot more room than a travel bag. Secure it to a seat, and give them food, hay, and water. You can allow them to ride on your lap if they are tame enough. Make sure to never separate a bonded pair. They need to remain together. Also, *never* leave them in a car unattended. The car will get hot quickly, and they can easily get heatstroke. During the winter, we will heat up their heat mat prior to leaving to make sure they are nice and warm. During the summer, we travel with a frozen water bottle to make sure they can keep cool. But never risk leaving them in the car. Once you get to your location, set up their area, and allow them to explore. They will probably have fun learning about a new place, and will eventually return to their area for a nap. Always remember other people's animals are not accustomed to your rabbit. It is your responsibility to keep them safe, and know what their limits are. For example, don't take them out on the beach on a hot day and stay for hours. Also, if there are a lot of dogs around, it will become far too stressful for them. However, once you figure out how to travel with your bunny, you both will have lots of fun adventures. The travel cage shown here is a full-size cage to allow him to move around and to have access to food, water, and the litter box.

Sugar loves to travel. He has no fear of the car, and little fear of anything, honestly, especially if we are nearby to protect him. He bounces around happily in his cage when we are traveling, and gets excited to see where we are when we arrive. He is tame enough to accomplish these trips easily now. Just remember, he is currently almost six years old, and has been traveling his entire life with us. He has gone to many places, and even quietly been in a travel bag that looks like a purse, and no one even realized he was with us.

Flying with Your Rabbit

There are many people who fly with their rabbits. Not all airlines allow this, so begin planning based on which airlines are okay with a rabbit flying with you, kept under your seat. You may need to provide documentation from your vet to establish health and good behavior. We would also recommend calling the airline and asking for a letter from them approving your rabbit to fly as a pet. Hopefully, this will help defuse any situation that may arise the day you are traveling. This is a decision you need to make based on your individual rabbit and their temperament. You will need to have food, hay, and water in your carry-on bag to make sure they are never going without the necessities. Also, prepurchase the pet ticket; don't assume it will be okay. In our opinion there are a lot of unknowns when it comes to air travel now that make it stressful and even dangerous for them. One important thing to note is that most airlines do not permit flying with rodents. This doesn't seem like a problem since we understand that rabbits are not rodents. However, security and flight attendants may not understand this, or believe you when you explain it to them. Never allow them to stow your rabbit in the animal section of the plane. This would be terrifying for them. If you need to fly with your rabbit, do everything you can to ensure the fact that the rabbit is allowed to be on the plane, and allowed to remain with you for the entire flight. Take their carrier with you, and make sure it will fit under the seat for takeoff and landing. Anything you can do prior to the flight to prevent any issues will be warranted. Also, you cannot stow a rabbit or any animal in the overhead luggage compartment. There is no air circulation there, and they will suffocate. If you are deciding to fly with your rabbit, just be prepared to not fly that day if something happens. Don't allow them to talk you into a situation that will put your rabbit in danger. Once your rabbit is no longer with you, you cannot control how well it is treated. This could end in a devastating loss that could have been avoided. Even if it costs you your ticket, know your rights, abide by the rules, but be ready to say, "Okay, I will cancel this flight and find another way." If you are in the right, the airline will probably reimburse you.

Leaving Your Rabbit Home

There are times when taking your rabbit with you isn't possible, and of course, if your rabbit hates to travel this happens more often. We always prefer to arrange for them to stay home in their own space, and in their own routine. We don't like to take them to a boarding place if we can avoid it. You have seen several mentions of what to do when you are away throughout this book, but we are going to compile a few of them here.

Hopefully, someone you and your rabbit know will agree to care for them while you are away. If you can afford a house sitter, it is even better, and usually cheaper than boarding. But, if you can't, having someone visit once a day will work. If you are having trouble locating someone to properly care for your rabbit, I suggest talking to your vet. There may be someone working at the vet's office who would be willing to do the daily visits just to make some extra money. They will be a great option because they will know how to handle your rabbit, as well as have access to the vet if something were to happen.

Make sure whoever is caring for your rabbit has been trained. They need to understand your rabbit's daily routine and maintain that as closely as possible. The best idea, I think, is to have a daily checklist for them, so that nothing is missed. If you aren't used to caring for a rabbit, you can easily forget something. Also, it is a good idea to have a back-up plan for your rabbit sitter. Your rabbit needs to be cared for daily, and if something happens and the person can't make it, someone else needs to check on them. Also, if they are on any type of medication, they need to be trained to administer that correctly.

Things to put on the list:

- Refill food bowls—put out fresh food bag.
- Clean and replace water bowls/bottles.
- Check litter box. Notice if there is more poop.
- Clean litter box if it is getting full.
- Make sure bedding is clean, replace if soiled.
- Make sure there is plenty of fresh hay out.
- Quick health check—Pooping? Pee? Happy or lethargic? Eating? Any odd behaviors that are different from other times?

Also be sure to include your preferred vet's name and number in case of emergencies and the location of your pet carrier, which should be ready to go.

It is best for you to plan your rabbits' food in advance so that they get the appropriate amounts and variety. There is a plan on page 65 for premade fresh food bags. Just create these bags based on what you would normally give your rabbit in a day. Also use small bags to put the correct amount of pellets for their bowl.

We like to leave a radio going softly for them, so it isn't so quiet. If there is a television in the room, you can leave that going sometimes. Our rabbits love to watch TV!

One last thing we try to do is have lots of toys ready for them to play with. We even make a few new ones that the rabbit sitter can offer to them daily.

Bunny Bits

If your rabbit is very attached to you, time away will be stressful and sad for them. If you normally wear a hoodie or bathrobe, consider removing the ties from it and leaving it on the floor for them. Being able to snuggle with a familiar article of your clothing, that smells like you, will be comforting for them. Just make sure there are no holes, strings, or ties that they could get tangled up in.

Boarding Your Rabbit

We personally do not like to board our rabbits. In our opinion, it is far too stressful for them. They are taken to a new place, probably confined to a smaller area, and they can hear all sorts of "predators" through the walls that will frighten them. However, if there is no other way to get them the care they need, just try to find a very nice place that gives them plenty of individual exercise time. Also, take the checklist for them—most boarding places rarely take care of rabbits. My fear is that if they aren't trained to handle them properly, your rabbit could easily get hurt, or stressed to the point of sickness. We are not saying there aren't great boarding places available, just not many with experience and a setup that is appropriate to small animals like rabbits. Just do research locally, and see if there is a nice place. Sometimes, there are very high-end veterinary practices that specialize in boarding rabbits. This would be a great option if you have one available. Visit this place prior to leaving your rabbit there, and ask them about their experience. You may have to drive a distance to find a great place, but believe us, it will be worth it to know your rabbit is safe and happy while you are away. This is why we believe arranging for them to remain at home in their own environment is the best for them.

Holidays

Holidays can be a fun and dangerous time for your rabbit! Our rabbits love Christmas, but you must take a few precautions. Electrical cords need to be covered and secured. Vacuum really well, and make sure to keep tree needles and tinsel off the floor. If you have a live tree, you will need to keep your rabbit away from it. Most live trees have been sprayed with chemicals that will be dangerous for them to ingest. Presents may become giant chew toys! Basically, for the first Christmas, keep a close eye and notice what intrigues them, and then *bunny proof*! But most importantly of all, make sure your rabbit has a great present under the tree so they feel included in the family fun!

Sugar dearly loves sleeping under the tree! He also likes to just stare at it. We have never had a problem with him wanting to eat it.

Rabbits on the Homestead

No need to worry here. In our opinion, rabbits are much more valuable than most people consider when it comes to the homestead. If you are raising rabbits as pets on your homestead, they can contribute so much! (We are not discussing meat rabbits or breeding rabbits in this book.)

Manure

Please don't tell us you simply toss your rabbit's litter waste!

Whether your rabbits are indoor or outdoor, their droppings are the best manure available. Let's have a little discussion about manure.

Generally speaking, all commercial grade fertilizers are made of N–P–K (Nitrogen–Phosphorus–Potassium). All plants need these three elements to survive, and rabbit manure is unique, because it has a very high level of nitrogen.

Many people ask what use a rabbit is on the farm if you don't intend on eating them. We are always perplexed by this question. What good is a cat then? Rabbits make great pets, and we advocate for keeping rabbits indoors as pets, and even on this small scale, a

single rabbit can make a vital contribution to your farm, or even your windowsill garden. Rabbit manure is also considered a "cold manure," which is another reason it is unique. This means that you can add it directly to your garden without needing it to go through the compost pile first. The tightly compacted droppings create a time-release fertilizer, because it takes several weeks for them to break down in the garden.

When we clean the litter box, we walk out to the garden, and simply dump it in the rows.

All winter this accumulates in the garden, and come spring, we put the tiller to it, and till it into the soil. Then, as we plant the garden, we still put the litter contents around the plants. Rabbit manure doesn't burn the plants at all, and therefore does not need time in the compost pile prior to being added to the garden. So, whether you are adding it to a garden, or putting it in a potted plant, make sure you make good use of your rabbit manure.

Fiber

Another item that your rabbit will give you is fur. Their beautiful fur can be spun and made into a beautiful thread. Longer haired breeds are preferable for this, but any rabbit can easily be brushed or carefully sheared and their fur saved for future use. Extremely short-haired breeds may not be suitable for spinning, but there are things you can do with their fur, too.

Most rabbit fur items that are on the market come from rabbits that are treated horribly in China. They are tortured and the fur is ripped from their skin. This is one reason why so many people are refusing to buy Angora items. Their beautiful fur comes at a very high price.

Beautiful Angora rabbits can be loved as valuable pets and members of your home and homestead. Giving them a happy life that is free from pain and cruelty is the goal. There are several varieties of Angora rabbits. Satin and French Angoras go through a shedding season about once every four months. During this time, their fur literally lets go, and you can easily groom them and get a good amount of fur that can be spun. German and English Angoras are very different. They tend to not shed their coats. In this case, you will need to carefully cut their fur. It is imperative that you take your time, and allow your rabbit to decide how long the grooming session will last. There are some good videos on YouTube to show you how to best trim your rabbit's fur. Here are a few

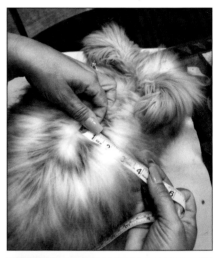

Photo Credit: Chrissy Morgan, Seventh Heaven Farm

basic guidelines, but please do your research and watch some good videos. There may even be some people local to you who would be willing to show you how.

1. Brush your rabbit carefully, removing any loose fur first.
2. Be prepared to take a couple days to fully trim your rabbit. You never want to stress your rabbit during a grooming session. They will remember it.
3. Rabbit fur is very fine, so make sure to have a pair of very sharp scissors that are dedicated to just trimming your rabbit.
4. Have cornstarch or some styptic powder on hand just in case.
5. Start at the back and gently begin to trim the fur parallel to the skin.
6. DO NOT PULL THE FUR TO TRIM. Any tension will increase the likelihood of you nicking their skin.
7. When your rabbit begins to show you they are tired, stop. Try again later, or on another day.
8. Be especially careful around their ears, tails, and folds in their skin. (People have actually accidentally cut off their rabbit's tail by not paying attention.)
9. I would suggest leaving any fur around their ears, tails, and feet. If at all, barely trim it. Do not get close to the skin.
10. When you are finished for the day, gently brush them again and use a blow dryer on a low setting to blow off any small pieces. You don't want them to ingest any pieces of fur that could stop up their GI tract.

A Rabbit Is Not a Rodent or a Hare

One of the major misconceptions about rabbits is that they are rodents.

Rabbits and hares are both classified as members of taxonomic order, the Lagomorpha. Specifically, they are both members of the family Leporidae within that order. They share very little in common with rodents, and it is very easy to distinguish the differences. However, even though they are a classified in the same family, rabbits and hares are not the same, either. Perhaps thinking about them like you would sheep and goats will make it easier to see how they can be classified as the same family and still be different.

A distinct difference between rabbits and rodents is their teeth. A rabbit has six incisors, whereas a rodent has two. A rabbit has 28 teeth that will never stop growing, while a rodent's incisors are the only teeth that continue to grow. The outer surface of a rodent's teeth is covered with enamel that is yellow, almost orange-ish in color, while a healthy rabbit's are white. A rabbit eats plants only, while a rodent eats plants and sometimes meat.

The most obvious difference between a rabbit and and a hare is that a rabbit's kits are born blind and hairless and a hare's kits are born with good vision and hair. A rabbit in nature will burrow underground, in a social situation, whereas a hare lives a more solitary life in a grassy nest aboveground. Hares are usually larger than rabbits, especially their ears and hind legs. Hares are not domesticated, while rabbits are more often bred as livestock and kept as pets.

Gardening for Your Bunnies

Having rabbits on your homestead is a great opportunity to expand your garden. They are giving you fertilizer for this garden, and so much of their daily feed can be generated from a garden that you are already growing. By planting another few plants, or even a separate garden, you can supplement their diet with food that you are growing already. Don't think that an extra garden will be a waste, because they will dearly love the fresh options, and they will devour them. If there are leftovers in the garden, that is perfect! Any chance that you get, pull the extra, and start drying it. This will be a great additive to their hay during the winter months. It is great to know that nothing is going to waste, and they will be able to taste the difference in their hay when you are sprinkling it with your freshly dried herbs. This will entice them to eat more of their hay, which will help keep their intestinal tract in great shape.

We try to grow a variety of foods specifically for the rabbits. We have a normal garden, but then we have a few places around the house that we are also growing herbs for them. It is so simple to add a few plants in the yard, in a container garden, or even in a window box. You will be shocked at the amount of extra that you can grow in a tiny space. We planted an entire box of kale and romaine that lasted for months. On the patio, we planted a container garden with several herbs in it, so that we could quickly pluck a few favorites if we needed a snack to give them. We also added this small raised bed that is totally dedicated to planting extras for them so that we can accumulate enough to dry for the winter.

We plant this garden fairly tightly so that we can make as much use of the space as possible.

Normally, we fill it with sage, oregano, thyme, lemon thyme, parsley, and lemon balm. We will plant sage off to the side because it grows so large, and we will also have echinacea growing in several other locations around the farm.

Final Notes

A Strong-Willed Bunny

We have four rabbits. If you can imagine, they are like children, each unique and different. Sugar is our angel. He wants to be with us and underfoot all the time. Chanel is sweet, and wants to be held. CoCo is aloof, and couldn't care any less about me unless we have food. Happy is our strong-willed bunny. If she can aggravate, she does her best, and yes, she knows what she is doing. She also hates being indoors.

For this reason, we are glad to admit that there is not one correct way to handle all rabbits. Just like with children, what works with one will not necessarily work with the next. All we have to do is say Sugar's name, and he knows he is being naughty, and tries to make up with us by snuggling with our feet. With Happy, that just inspires her to be naughtier. It is like she says, "Oh that bothers you? Good. Watch this." We will admit, we have learned more from attempting to train Happy than we have learned from training our other three. Honestly, training the others was so easy because they basically trained themselves. They are smart, treat seeking, and very neat. Happy, on the other hand, wants everything her way. She will not litter box train; she prefers a pee pad. It probably took us three months to say, "Okay, you win." We purchased reusable pee pads, and she won. It wasn't worth the constant fight and irritation. She had apparently been raised with a pee pad prior to being abandoned, and she probably thought we were so mean trying to make her use a box. We bought pads, tried different boxes, different litters, and we even put pee pads in the boxes, but nothing worked. She wasn't peeing all over the floor. She never missed, and she wasn't going to give up. She used a pee pad. Once we resigned ourselves to this, things got better. Was she perfect? No, but better. We weren't expecting her to do what we wanted, and that helped our relationship. On the other hand, she does not chew wood at all. She has never messed anything up. Our other little chewers have managed to get the occasional door frame or furniture leg if we aren't paying attention. Happy prefers an old towel. She loves to shred them.

You have to learn what your rabbit responds to, and then adjust your expectations. Never try physical discipline—it's too dangerous, and you risk serious injury. You can do time-outs and a stern voice, but that is the limit of discipline. Positive reinforcement is the only thing they seem to respond to.

Happy went through a stage where she thought biting was okay. Almost everything will correct itself with enough time. As they age past the 18-month mark, most of these behaviors will slow down and go away. You just have to have patience, and be consistent.

What to Do If You Just Can't Do It

This is the sad part of rabbit keeping. People often purchase these adorable animals on a whim. They aren't prepared for the commitment it takes to provide a happy home for them. For many reasons, people decide that they can't keep their rabbits. This is especially sad for the misunderstood rabbit. However, there are good things to do when attempting to re-home a rabbit, and some very bad things to do. Here are a few things to consider before making a decision.

ABANDONMENT

Your initial thought may be to simply turn them loose. Whether you do this in your yard, a park, or the woods, it is a death sentence. Domestic rabbits do not generally survive in the wild. They do not know how to source safe foods or water like wild rabbits do, and diseases that wild rabbits carry are many times fatal to domestic rabbits. In addition to this, rabbits are very territorial, so abandoning your rabbit into another rabbit's territory will cause a fight that the domestic rabbit doesn't know how to handle. Finally, predators will more than likely find your rabbit quickly.

SHELTERS

Shelters are better than abandonment sometimes. Please check any shelter prior to dropping off a rabbit. Are they a no-kill shelter? Are they overfilled already? Most shelters will only keep rabbits for a few days, if they will accept them at all. If they aren't adopted or rescued quickly, they will be euthanized. If, in their fear of the new, scary environment, they show any aggression, they may not make it even a few days.

INTERNET ADS

Offering up your rabbit for free on an Internet site is another really bad idea. There are plenty of people willing to take a rabbit for many reasons other than providing a loving home, which may be what they tell you. People take rabbits all the time to cook, feed to other pets, or even to train dogs to hunt.

RE-HOME

Your rabbit's best chance at survival will be to locate a loving family to take them, or locate a local rabbit rescue who is prepared to handle the situation. If you don't know of a place, there are lots of rabbit groups on Facebook that can help you rehome, and you can always google the closest rabbit or small animal rescue. Perhaps even a local farm supply store can be asked if they have any ideas, or if they know of anyone who would be willing to take a rabbit.

A Friend for Life

We hope we have helped you decide and prepare to get a rabbit of your very own. They are such a source of joy, friendship, entertainment, and love. They will enrich your life! When you adopt your new friend, please consider the things you have read here, and know that you will get as much out of this relationship with your rabbit as you are willing to put into it. Patience will be key, but if you will take the time necessary to bond with your rabbit and train them, you will enjoy great success! Always remember that each rabbit will have a unique personality. You will never have two rabbits that are exactly the same, just as you will never find two people who are exactly the same. Your rabbit will eventually tightly bond to you, and if they are willing to love you for their entire life, the least you can do is give them the very best that you possibly can. Always remember to keep your expectations of them in check. They are not a dog that will worship you from day one, they are not a cat you will necessarily be able to cuddle immediately—you have a rabbit whose trust you will have to earn slowly. All you really have to do is love them.

Acknowledgments

We would like to take a moment to thank a couple of people who greatly helped us during the writing process of this book. While we wrote this book from our experiences and research, we are self-taught.

To our delight, Dr. Micah Kohles helped answer many questions and gave his educated opinions on many items throughout this book. In addition to his help, Janet Garman read through the manuscript twice helping to edit, and she also contributed photographs from her farm.

Micah Kohles, DVM, MPA is the Co-Founder of Flatwater Veterinary Group, where he works with a large diversity of exotic species, as well as Vice President of Technical Services and Research for Oxbow Animal Health. Dr. Kohles serves as adjunct professor in the School of Veterinary Medicine and Biomedical Sciences at the University of Nebraska – Lincoln and is the past President of the Association of Exotic Mammal Veterinarians (AEMV). He lectures regularly at conferences, symposiums, and pet owner events, helping educate veterinary professionals and pet parents alike on the proper nutrition and care of small animals.

Janet Garman is a farmer and freelance writer. She has a degree in Animal Science from the University of Maryland and has raised many species of animals both on her farm and in her home. *50 Do-It-Yourself Projects for Keeping Chickens* (Skyhorse Publishing, 2018) is her latest book. Janet's other books include *Chickens from Scratch: Raising Your Own Chickens from Hatch to Egg Laying and Beyond*; *Habitat Housing for Rabbits*; and *Margarita and the Beautiful Gifts*. She lives in Crownsville, Maryland.

Resources

Favorite Brands:
Oxbow: Pellets, Hay, Treats, Critical Care
VetRx: Natural Respiratory Remedy
Vetericyn: Wound Care, Eyewash, Opthamalic Gel

Suggested Websites for Additional Research
Happy Days Farm: www.happy-days-farm.com
Oxbow Animal Health: www.oxbowanimalhealth.com
House Rabbit Society: www.rabbit.org

Additional Resources and References
Craven, Boyd, and Rick Worden. 2013. *Beyond the Pellet: Feeding Rabbits Naturally.*
Moore, Lucile, and Kathy Smith. 2008. *When Your Rabbit Needs Special Care: Traditional and Alternative Healing Methods.*
Vintage Farming Classics: The Health & Diseases of Rabbits: A Collection of Articles on the Diagnosis and Treatment of Ailments Affecting Rabbits.

Index

second-cut hay, 51
seeds, 68
senses, 35–40
shelters, 137
shock, 119
sight, 36
simethicone, 102
sleep, 73–74
smell, sense of, 37–38
smoothing, 95
snuffles, 106
sore hock, 46, 115
spay, 10, 16, 34, 41–42
spinach, 67
splay leg, 121
spraying, 100
spring greens, 67
squash, 67
stairs, 78–80
stethoscope, 102
styptic powder, 102
surgery, 103–105
Sussex, 9

T
taste, sense of, 38–39
teeth, 40–41, 134
thumping, 99
thyme, 59, 67
timothy hay, 30, 50
tomato, 67
tooth spurs, 40
touch, sense of, 39–40
toxic foods, 68
toys, 91–94
training, litter box, 81–83

travel
    boarding during, 129
    by car, 123–126
    leaving at home during, 94, 127–128
    by plane, 126–127
treats, 66–68
trimming nails, 87–88
turnip greens, 67

U
underweight, 72
urinary tract infections, 120–121

V
vegetables, 67
Vetericyn Eyewash, 103
Vetericyn Ophthalmic Gel, 103, 122
Vetericyn Wound Care, 102
veterinarian, 19–20
Vet Rx Oil for Rabbits, 101
Vet Wrap, 103
vision, 36

W
warbles, 116
water, 47–49, 77
watercress, 67
weepy eye, 122
wheatgrass, 67
wood, 25–26, 91
wound care, 102

Z
zucchini, 67